The aim of the Handbooks in Practical Animal Cell Biology is to provide practical workbooks for those involved in primary cell culture. Each volume addresses a different cell lineage, and contains an introductory section followed by individual chapters on the culture of specific differentiated cell types. The authors of each chapter are leading researchers in their fields and use their first-hand experience to present reliable techniques in a clear and thorough manner.

Endothelial Cell Culture contains chapters on endothelial cells derived from lung, bone marrow, brain, mammary glands, skin, adipose tissue, female reproductive system and synovium.

Endothelial cell culture

Endothelial cell culture

Edited by

Roy Bicknell

Molecular Angiogenesis Group,
Imperial Cancer Research Fund,
Institute of Molecular Medicine,
University of Oxford, Oxford UK

Published by the Press Syndicate of the University of Cambridge
The Pitt Building, Trumpington Street, Cambridge CB2 1RP
40 West 20th Street, New York, NY 10011-4211, USA
10 Stamford Road, Oakleigh, Melbourne 3166, Australia

First published 1996

Printed in Great Britain at the University Press, Cambridge

A catalogue record for this book is available from the British Library

Library of Congress cataloguing in publication data

Endothelial cell culture / edited by Roy Bicknell
p. cm. – (Handbooks in practical animal cell biology)
Includes index.
ISBN 0 521 55024 6 (hardcover : alk. paper). – ISBN 0 521 55990 1 (pbk. : alk. paper)
1. Endothelium – Cultures and culture media – Laboratory manuals.
2. Cell culture – Laboratory manuals. I. Bicknell, R. J. (R. John)
II. Series.
QM562.E53 1996
611'.0187–dc20 96-5699 CIP

ISBN 0 521 55024 6 hardback
ISBN 0 521 55990 1 paperback

SE

Contents

Contributors

Stewart E. Abbot
Guy's Renal Laboratories, Guy's Tower, Guy's Hospital, London, SE1 9RT, UK

Roy Bicknell
Molecular Angiogenesis Group, Imperial Cancer Research Fund, Institute of Molecular Medicine, University of Oxford, Oxford, OX3 9DU, UK

David R. Blake
The Bone and Joint Research Unit, 25-29 Ashfield Street, London, E1 2AD, UK

William W. Carley
Bayer Research Center, 400 Morgan Lane, West Haven, CT 06516, USA

Peter W. Hewett
University of Nottingham, Laboratory of Molecular Oncology, CRC Department of Clinical Oncology, City Hospital, Hucknall Road, Nottingham, NG5 1PB, UK

Lisa Masek
CRC Wessex Medical Oncology Unit, University of Southampton, Southampton General Hospital, Tremona Road, Southampton, SO16 6YD, UK

J. Clifford Murray
University of Nottingham, Laboratory of Molecular Oncology, CRC Department of Clinical Oncology, City Hospital, Hucknall Road, Nottingham, NG5 1PB, UK

Margaret C. P. Rees
Nuffield Department of Obstetrics & Gynaecology, John Radcliffe Hospital, University of Oxford, Oxford, OX3 9DU, UK

Zbigniew Ruszczak
Dermatology, Dermatopathology and Experimental Dermatology Visiting Consultant, Department of Dermatology, New Jersey Medical School
Correspondence to: Koenigstrasse 246, D-32427 Minden, Germany

Charlotte Schulze

EISAI London Research Laboratory, Bernard Katz Building, University College London, Gower Street, London, WC1E 6BT, UK

Clifford R. Stevens

The Bone and Joint Research Unit, 25–29 Ashfield Street, London, E1 2AD, UK

John Sweetenham

CRC Wessex Medical Oncology Unit, University of Southampton, Southampton General Hospital, Tremona Road, Southampton, SO16 6YD, UK

Stuart K. Williams

Section of Surgical Research, Department of Surgery, University of Arizona Health Sciences Center, Tucson, AZ 85718, USA

Yuan Zhao

Nuffield Department of Obstetrics & Gynaecology, John Radcliffe Hospital, University of Oxford, Oxford, OX3 9DU, UK

Preface to the series

The series Handbooks in Practical Animal Cell Biology was born out of a wish to provide scientists turning to cell biology, to answer specific biological questions, the same scope as those turning to molecular biology as a tool. Look on any molecular cell biology laboratory's bookshelf and you will find one or more multivolume works that provide excellent recipe books for molecular techniques. Practical cell biology normally has a much lower profile, usually with a few more theoretical books on the cell types being studied in that laboratory.

The aim of this series, then, is to provide a multivolume, recipe-book-style approach to cell biology. Individuals may wish to acquire one or more volumes for personal use. Laboratories are likely to find the whole series a valuable addition to the 'in house' technique base.

There is no doubt that a competent molecular cell biologist will need 'green fingers' and patience to succeed in the culture of many primary cell types. However, with our increasing knowledge of the molecular explanation for many complex biological processes, the need to study differentiated cell lineages *in vitro* becomes ever more fundamental to many research programmes. Many of the more tedious elements in cell biology have become less onerous due to the commercial availability of most reagents. Further, the element of 'witchcraft' involved in success in culturing particular primary cells has diminished as more individuals are successful. The chapters in each volume of the series are written by experts in the culture of each cell type. The specific aim of the series is to share that technical expertise with others. We, the editors and authors, wish you every success in achieving this.

ANN HARRIS

Oxford,
July 1995

Acknowledgement

The editor thanks Dr Annabel Smith, Institute of Molecular Medicine, for her expert preparation of the index.

1

Introduction to the endothelial cell

Roy Bicknell

Introduction

For many years the endothelial cell was regarded in medicine and physiology as 'an inert sac whose function was simply to separate the blood from the tissues'. We now know that the endothelium is an extremely active tissue that plays an intimate role in many physiological processes. Dysfunction of the endothelial cell is a key step in the development of many pathologies including the two major killers of Western man: atherosclerosis (a disease of the larger vessels) and cancer (tumour angiogenesis). As the skill of biologists in the characterisation of the endothelium *in vivo* and *in vitro* has increased it has become apparent that there exists extensive endothelial heterogeneity. Thus, the tenet that an endothelial cell is an endothelial cell is an endothelial cell is no longer valid. A corollary of this heterogeneity is that it is imperative to work with an appropriate endothelial isolate when studying the role of endothelium in a given pathology. From this has sprung the increasing interest in the isolation and culture of endothelium from diverse organs.

A remarkable property of endothelium is its quiescence. Thus, in the adult the average endothelial cell divides (approximately) twice in a lifetime. Contrast this with epithelia that divide continually throughout adult life. Nevertheless, when required, endothelial proliferation can be very rapid. Consider a healing wound: endothelial proliferation (as a component of angiogenesis) is switched on, then off, as required and is under exquisite control. In the past I have likened this angiogenesis to the blood clotting cascade. The existing vasculature is a coiled spring poised for angiogenesis should a physiological need arise.

The quiescence of endothelium *in vivo* is a situation that we would like to mimic *in vitro*. Cells may be left unfed to induce quiescence; however, endothelial biologists have sought *in vitro* models that are closer than this to the *in vivo* state of the cell. One of the most frequently used techniques is to seed endothelial cells onto the extracellular matrix mimic Matrigel (an extract of the Engelbreth–Holm–Swarm murine tumour). On this matrix the cells form so-called capillary tubes. That seeding on Matrigel does indeed push the cells down the differentiation pathway is supported by the observation that bovine adrenal capillary endothelial cells strongly express CD31 *in vivo* but not *in vitro*, unless they are seeded onto this matrix (Fawcett, 1994). There remains debate as to whether the tubes formed on Matrigel possess a lumen; I remain sceptical from my own experience. More convincing 'tubes' are seen when the cells are seeded into a two-dimensional matrix of collagen; unfortunately quantitation is then a greater problem.

The purpose of this volume is to provide a practical guide for those interested in the isolation and culture of endothelium from a chosen organ. The organs covered in this volume have been selected by two criteria: (i) their pertinence to areas of currently active medical research and (ii) that they are organs for which significant advances in the isolation and culture of the endothelium have been reported. The selection is not inclusive. Organs not covered here but worthy of mention include the adrenal medulla, for which it has long been known that the sinusoidal endothelium is peculiarly amenable to isolation by clonal selection and long-term culture (Folkman *et al.*, 1979), and the liver, for which convincing, and comparatively early, papers described the isolation of the sinusoidal endothelium by elutriation (Irving *et al.*, 1984; Shaw *et al.*, 1984). Nevertheless, for many organs, such as the colon or placenta, it is true to say that little progress, if any, in the isolation and culture of the microvascular endothelium has yet been made. [*Note added in proof:* No longer true for colon. See Haraldson *et al.* (1995).]

Large vessel versus microvascular endothelium

The endothelium covered in this volume is all microvascular with one exception, in that a protocol is given for the isolation of human umbilical endothelial cells (or HUVEC) in Chapter 3. The reason for the bias to microvascular endothelium is that the isolation of large vessel endothelium by either cannulation followed by perfusion of the vessel with collagenase and flushing out of the cells (as for HUVEC) or by opening up of the vessel and removal of the endothelial monolayer by gentle scraping, are so easy and routine that their incorporation in this volume is not justified. Further, large vessel endothelium is far less fastidious in its growth requirements and may usually be cultured on standard polylysine coated plasticware in 10% fetal calf serum without other supplements (HUVEC are an exception, requiring gelatin coated plasticware and 'endothelial cell growth supplement' or acidic or basic fibroblast growth factor).

Some other comments concerning the HUVEC are justified. Very widely used (because of free access to tissue and ease of isolation) the HUVEC is nevertheless a unique endothelium exhibiting properties that are in many ways intermediate between those of large vessel (e.g. aortic) and microvascular endothelium. Thus, growth supplements not present in serum and gelatin or fibronectin coating of plasticware are common requirements for culture of microvascular but not large vessel endothelium. The rapid loss of endothelial characteristics in culture (HUVEC should never be used beyond the fourth passage) is unique to this endothelium and probably reflects a terminal programming. [This has been shown to be the result of production of interleukin (IL)-1α by the HUVEC itself. Administration of antisense IL-1α oligomers blocks senescence (Maier *et al.*, 1990).] Despite this, the HUVEC has, to the delight of those interested in inflammation, diapedesis and cell adhesion, proved to be an excellent model of the post-capillary venule. It is known that diapedesis *in vivo* occurs in the post-capillary venule; however, endothelium from that site has proved difficult to isolate and culture. The rat lymph node post-capillary venule endothelial cell synthesises a unique sulphated glycolipid that it incorporates into the plasma membrane. This sulphated glycolipid provides a means with which to track the venular endothelial cell during isolation after *in vivo* labelling with radioactive sulphate (Ager, 1987). Bovine lymphatic endothelium has also been isolated, characterised and cultured (see Pepper *et al.* 1994).

Immortalised endothelial cell lines

Despite many attempts to immortalise endothelial cells from various sources these have, by and large, been somewhat disappointing. The principal reason for this is that upon immortalisation endothelial cells frequently lose their endothelial characteristics and a cell line is obtained that shows greater similarity to a fibroblast than an endothelial cell. Cell biologists not surprisingly screen their immortalised clones for lines that exhibit the phenotypic characteristics they are interested in. While such a line is often found, it almost inevitably has lost other characteristics of the primary isolate. This is not to say that such cell lines are not of use; they clearly are in some studies. There have been claims of success in obtaining endothelial lines that exhibit a good repertoire of endothelial characteristics and these are worthy of note. Immortalised HUVEC lines have been described by researchers at British

Biotechnology (Green *et al.*, 1994). These workers used SV40 large T antigen under control of either the Rous sarcoma virus long terminal repeat or the human cytomegalovirus promoter. They were interested in obtaining lines that retained cytokine induction of adhesion molecules (e.g. E-selectin and vascular cell adhesion molecule, VCAM). An immortalised human skin microvascular endothelial cell line known as HMEC-1 has also been reported (Ades *et al.*, 1992).

We have for some time worked with the immortalised mouse endothelial lines known as SEND and BEND. Engineering of a transgenic mouse expressing polyoma middle T antigen under control of the viral thymidine kinase promoter gave rise to a strain of mice that developed spontaneous haemangiomas (Williams *et al.*, 1988). The isolation of endothelium from these mice proved quite straightforward and the isolates were found to be immortal, giving rise to SEND (from skin) and BEND (from brain) (Williams *et al.*, 1988, 1989) These lines are exceptionally easy to culture and retain many useful properties, not least of which is inflammatory cytokine induction of the adhesion molecules E-selectin and VCAM. It would be interesting to examine immortalisation of human endothelial cells with polyoma middle T rather than the commonly used SV40 large T that frequently results in de-differentiated lines.

Commercially available endothelial isolates

The retailing of primary endothelial isolates has been poineered by the Clonetics Corporation of California (their British distributor is TCS Biologicals Ltd). Clonetics first started marketing HUVEC some years ago and quickly gained a reputation for the sale of a quality product. In view of the fact that HUVEC are good only for four passages and because HUVEC are so easy to isolate, provided that a ready supply of umbilici is at hand, any laboratory requiring regular supplies would probably not be interested in their purchase. In contrast, if a laboratory required HUVEC for short-term experiments, or if financial considerations were not a limiting factor, then they are undoubtedly a good buy. Since the initial sales of HUVEC, Clonetics have moved on to market human dermal microvascular endothelial cells (HDMEC: see Chapter 6). We have found these isolates to be excellent. HDMEC are sold at passage 1 and are readily cultured in our hands for at least eight passages without loss of cobblestone morphology or endothelial characteristics. In view of how difficult these cells are to isolate, purchase from Clonetics provides an excellent alternative. Recently Clonetics have started marketing human lung microvascular endothelial cells.

References

Ades, E.W., Candal, F.J., Swerlick, R.A., George, V.G., Summers, S., Bosse, D.C. & Lawley, T.J. (1992). HMEC-1: establishment of an immortalized human microvascular endothelial cell line. *J. Investi. Dermatol*, **99**, 683–90.

Ager, A. (1987). Isolation and culture of high endothelial cells from rat lymph nodes. *J. Cell Sci.*, **87**, 133–44.

Fawcett, J. (1994). Molecular aspects of tumour angiogenesis and metastasis. D.Phil thesis, University of Oxford, Oxford.

Folkman, J., Haudenschild, C.C. & Zetter, B.R. (1979). Long-term culture of capillary endothelial cells. *Proc. Natl. Acad. Sci. USA*, **76**, 5217–21.

Green, D.R., Banuls, M.P., Gearing, A.J.H., Needham, L.A., White, M.R.H. & Clements, J.M. (1994). Generation of human umbilical vein endothelial cell lines which maintain their differentiated phenotype. *Endothelium*, **2**, 191–201.

Haraldson, G., Rugtveit, J., Kvale, D., Scholz, T., Muller, W.A., Hovig, T. & Brandtzaeg, P. (1995). Isolation and longterm culture of human intestinal microvascular endothelial cells. *Gut*, **37**, 225–34.

Irving, M.G., Roll, F.J., Huang, S. & Bissel, D.M. (1984). Characterization and culture of sinusoidal endothelium from normal rat liver: lipoprotein uptake and collagen phenotype. *Gastroenterology*, **87**, 1233–47.

Maier, J.A.M., Voulalas, P., Roedner, D. & Maciag, T. (1990). Extension of the life-span of human endothelial cells by an interleukin-1α antisense oligomer. *Science*, **249**, 1570–4.

Pepper, M.S., Wasi, S., Ferrara, N., Orci, L. & Montesano, R. (1994). *In vitro* angiogenic and proteolytic properties of bovine lymphatic endothelial cells. *Exp. Cell. Res.*, **210**, 298–305.

Shaw, R.G., Johnson, A.R., Schulz, W.W., Zahlten, R.N. & Combes, B. (1984). Sinusoidal endothelial cells from normal guinea pig liver: isolation, culture and characterization. *Hepatology*, **4**, 591–602.

Williams, R.L., Courtneidge, S.A. & Wagner, E.F. (1988). Embryonic lethalities and endothelial tumours in chimeric mice expressing polyoma virus middle T oncogene. *Cell*, **52**, 121–31.

Williams, R.L., Risau, W., Zerwes, H.G., Drexler, H., Aguzzi, A. & Wagner, E.F. (1989). Endothelioma cells expressing the polyoma middle T oncogene induce hemangiomas by host cell recruitment. *Cell*, **57**, 1053–63.

2

Lung microvascular endothelial cells: defining *in vitro* models

William W. Carley

Introduction and application

The tremendous vascular surface area represented by the pulmonary endothelium aids in essential blood/gas exchange but also effectively regulates circulating levels of vasoactive agents. For example, protease activities on the endothelial surface can dictate total blood levels of critical vasoregulatory peptides such as angiotensin converting enzyme (ACE). The continuous capillaries of the lung comprise the majority of this surface area and maintain a highly selective permeability barrier that, when compromised, can lead to oedema and pulmonary pathology.

The microvessels of the lung undoubtedly develop some of their physiological activities from interactions with other pulmonary cells. Information exchange with epithelial type I and type II cells occurs at the level of secreted and extracellular matrix proteins to dictate proper airway architecture and vascular function. It is the understanding at this cellular level that has driven researchers to take the simplistic approach of isolating cell types and culturing them *in vitro.* But how much information do isolated and repeatedly cultured cells 'remember' about how to mimic their functions *in situ*? If this question is asked frequently by cell and molecular biologists who desire the optimal *in vitro* model of a vascular bed, our understanding of whole-organ physiology will consequently benefit.

The application of *in vitro* models of the pulmonary microvasculature must be tempered with defining the cell type as well as the desired physiological measurement. For example, the use of ACE as a reliable marker for pulmonary endothelium was aided by the knowledge that angiotensin could be cleaved when passed through an isolated lung preparation. This may indicate that a unique result in an animal pulmonary model may invite the isolation of the microvessels from that animal with the desire to measure an endothelium-specific enzyme. However, culturing techniques have improved so effectively that isolating specific cells from an entire organ now demands specific markers for these cell types. In general, present-day endothelial cell markers are in most endothelial cells and have not been refined to the point of identifying a specific cell type from a particular vascular bed. Combinations of markers, as we will see, may be able to refine this approach. Eventually, molecular cloning techniques may help identify such novel markers, but until then, reliable isolation methods, cell sources, characteristics, long-term culture conditions and cryopreservation techniques must be the tools of *in vitro* discovery research on the endothelium. Since the isolation of pulmonary endothelium from different species results in variability in many of these tools, different species isolates of the pulmonary vasculature will be described under each section. The cell characteristics and markers to be described will be only those used with pulmonary microvessel endothelial cells in culture. A rule of thumb used in our laboratory for markers is always to compare expression of a marker in cells and tissue. Often, however, because antibodies directed against markers do not cross species barriers, cell-morphology selection must, in some cases, be used.

Rodent pulmonary microvascular endothelial cells

Initial isolates of rat pulmonary microvessel endothelium (RPME) employed the retrograde perfusion of collagenase in intact lungs to dislodge the resident endothelium (Habliston *et al.*, 1979), but no markers existed to determine whether large or small vessel cells predominated in such cultures. In a novel perfusion approach, Ryan *et al.* (1982) non-ezymatically disrupted the vasculature of the lung with cold EDTA and perfused (antegrade and retrograde) beads of 40–80 μm diameter to which pre- and post-capillary endothelium could attach and thereby be isolated. However, the largest cell type contributing to lung physiology was, by definition, absent: the capillary cells. These early techniques often also required perfusable lung tissue for isolation rather than a small segment. More recently partial peripheral segments of lung have been used for homogenates that contain a mixture of cell types, but are predominantly microvessel endothelium (Magee *et al.*, 1994). Without using a molecular marker to select cell type, a physical filter separation was sufficient to yield a homogeneous population that carried four specific molecular markers. The technique of Magee, *et al.* (1994) [based on Chung-Welch *et al.* (1988) and Davies *et al.* (1987)] is detailed below because it is relatively straightforward and results in a mixture of microvessel cells which are, assuming no short-term growth advantages between cell types, mostly capillaries.

Chemicals and reagents

Primaria tissue cultureware (Becton-Dickinson, Lincoln Park, NJ)
Equine plasma-derived serum (PDS; Cocalico Biologicals, Reamstown, PA)
Defined fetal bovine serum (FBS; Hyclone Laboratories, Logan, UT)
Buffered salt solutions, media, non-essential amino acids (NEAA) and 0.5% trypsin-EDTA (GIBCO, Grand Island, NY)
L-Glutamine (JRH Biosciences, Lenexa, KS)
Bovine serum albumin (BSA) fraction V, gelatin, antibiotic–antimycotic solution, heparin, endothelial cell growth factor (ECGF) and D-valine (Sigma, St Louis, MO)
Type II collagenase (Worthington Biochemical, Freehold, NJ)
Nylon mesh (20, 100 μm; Tetko, Lenexa, KS)

Cell materials

1,1′Dioctadecyl-3,3,3′,3′-tetramethyl-indocarbocyanine perchlorate acetylated [DiI-Ac]-low density lipoprotein (LDL) or DiI-Ac-LDL (Biomedical Technologies, Stoughton, MA)
Fluorescein isothiocyanate (FITC)-conjugated goat anti-mouse polyvalent immunoglobulin G (IgG) and FITC-conjugated *Bandeiraea simplicifolia* I isolectin B_4 (BSI) (Sigma, St Louis, MO)
MRC OX-43 ascites (IgG1; Harlan Bioproducts, Madison, WI)
RECA-1 (provided by Dr A. M. Duijvestijn, Dept of Immunology, U. Limburg, Maastricht, The Netherlands)

Protocol for rat lung microvessel cell isolation and cell growth

1 After exsanguination (anaesthetic: ketamine HCl 100 mg/kg i.m.), remove the lungs from Sprague–Dawley rats (pathogen-free, males, 150–200 g; Charles River, Raleigh, NC) (At this point submersion of the lungs in 70% ethanol in water will devitalise the mesothelial layer.)
2 Remove the visceral pleura and the outer 3–5 mm of peripheral lung tissue and 'finely' mince the remaining pieces. Wash the pooled pieces with Hank's Balanced Salt Solution (HBSS) and collect on a 20 μm nylon filter mesh.
3 Digest the tissue with constant mechanical agitation with 0.5% collagenase (in calcium-, magnesium-free HBSS containing 0.5% BSA) at 25 °C for 1 h. (Note: Although the authors do not give a ratio of digestion buffer

volume to tissue volume, our experience has shown a ratio of 5:1 is the minimum for effective digestion.)

4 Vortex the suspension and wash twice by centrifugation (200 *g*, 5 min), and resuspend in the above collagenase buffer without BSA (maintaining the same volume ratio).
5 Centrifuge again and resuspend the pellet in plating medium (Dulbecco's Modified Eagle Medium (DMEM), 10% PDS , 90 μg/ml heparin, 150 μg/ml ECGF, 2 mM L-glutamine and 1% antibiotic–antimycotic solution).
6 Pass the mixture through 100 μm nylon mesh, plate in one 60 mm Primaria plate and grow in a humidified incubator at 37 °C at 10% CO_2.
7 Re-feed cells after 2 h with plating medium to remove non-adherent cells.
8 After 48 h re-feed cells with L-valine-free medium (L-valine-free MEM, 15% FBS, 90 μg/ml heparin, 60 μg/ml ECGF, 4 mM L-glutamine, 0.1 mM NEAA, 1.2 g/l D-valine and 1% antibiotic–antimycotic solution) and culture in a humidified incubator at 37 °C in 5% CO_2.
9 Passage cells at confluence at a 1:3 split using trypsin/EDTA.
10 Maintain cells in this medium for three passages and freeze (The authors do not give a medium for freezing but our best success with any cell line is with using 10% dimethylsulphoxide (DMSO) and 90% FBS.)

The cells were characterised by the uptake of DiI-Ac-LDL, the binding of the lectin BSI and by labelling with monoclonal antibody MRC OX-43 (rat endothelium-specific – except brain) and RECA-1 (rat endothelium-specific). (The experimenter is referred to the original paper for their conditions for staining the cells and the original papers on the monoclonal antibodies: see Ryan *et al.*, 1982.)

Other sources of reagents used to characterise rat pulmonary microvessel endothelium from a variety of sources are:

Tube formation: Matrigel (Collaborative Research, Bedford, MA)
Factor VIII related antigen (Behring Diagnostics, American Hoechst Corp., Sommerville, NJ)

Protocol for mouse lung microvessel cell isolation and growth

As a refinement to the general techniques described for different species isolation of pulmonary microvessel cells, the reader is referred to a recent work on mouse lung microvessel cell isolation by Gerritsen *et al.* (1995). The entire, somewhat intricate technique will not be described here but the protocol follows steps of collagenase perfusion, mincing, further digestion, homogenising, filtering and centrifugation prior to fluorescence-activated cell sorting (FACS) to enrich for microvessels. The strength of the FACS approach (assuming accessibility to such a facility) was demonstrated by sorting the cells for vascular cell adhesion molecule 1 (VCAM-1) expression after activation by lipopolysaccharide (LPS) administration to the mouse prior to sacrifice. After 1–2 weeks of growth, the cells were sorted for DiI-Ac-LDL uptake – a double marker sort. The resulting population expressed von Willebrand factor (vWF), platelet–endothelial cell adhesion molecule (PECAM), thrombomodulin, lung-endothelial cell adhesion molecule (LuECAM), constitutively, and intercellular adhesion molecule 1 (ICAM-1), VCAM-1 and E-selectin after interleukin 1β (IL-1β) stimulation. A comparison of markers between the cells and the inflamed tissue concluded that the population includes small collecting veins, venules, septal capillaries and small arteries. Since the characterisation of these cells was so thorough and since the reagents successful in mice may also be useful in rats, the sources of the antibodies used in this work should be mentioned:

Anti-human von Willebrand factor (vWF; DAKO, Carpineria, CA)
Anti-murine PECAM (Dr S. Bogen, Boston University, Boston, MA)
Anti-murine thrombomodulin (Dr S.J. Kennel, Oakridge National Laboratory, Oakridge, TN)
Anti-LuECAM (Dr B. Pauli, Cornell University, Ithaca, NY)
FITC-conjugated *Bandeiraea simplicifolia* agglutinin (BSI B_4; E-Y Laboratories, San Mateo, CA)
Anti-murine ICAM-1 (YN/1; American Type Culture Collection[ATCC])
Anti-VCAM-1 (MK 1.91; ATCC)
Anti-murine E-selectin (9A9; Dr B. Wolitsky, Hoffman La Roche, Nutley, NJ)

The cells are grown in DME/F12 (GIBCO, Grand Island, NY), 20% FBS (Hyclone, Logan, UT), 6% PDS (Sigma, St Louis, MO), 2 mM L-glutamine, 2 μl/ml RDGF,* 90 μg/ml heparin (Fischer, New Brunswick, NJ).

Bovine pulmonary microvascular endothelial cells

One of the most widely used and adapted protocols for bovine microvessel endothelial cells has been that of Chung-Welch *et al.* (1988). This technique

* Retinas were removed from bovine eyes and incubated for 3 h at 25 °C in HBSS (1 retina/1 ml). The mixture was centrifuged (600 *g*) for 5–10 min at 25 °C. The supernatant was passed through a 0.2 μm filter (Nalgene, Rochester, NY)

was adapted from that described above for rat pulmonary microvessel cells and others and uses the size of the organ to the advantage of the researcher. Since the bovine lung is large the peripheral lung sections used can be selected to be free of large vessels and contaminating pleura and mesothelial cells.

Another similar technique, by Del Vecchio *et al.* (1992), based on that of Folkman *et al.* (1979), uses a cloning ring which isolates the endothelial cells on the basis of morphology. As a result, the technique to be described, that of Chung-Welch *et al.* (1988), will probably produce a broader spectrum of microvessel endothelium since it does not involve this cloning step. The researcher may desire further cell fractionations after this initial isolation (see the protocol for mouse lung microvessel cell isolation, p. 10, or the Dynabead protocol, p. 18).

Protocol for bovine microvessel endothelial cell isolation and growth

1 Obtain whole lungs from 2- to 3-week-old calves from an abattoir and transport on ice to the laboratory. (At this point submersion of the lungs in 70% ethanol/water will devitalise the mesothelial layer.)
2 Remove the visceral pleura and the outer 3–5 mm of peripheral lung tissue and 'finely' mince the remaining pieces. Wash the pooled pieces with Dulbecco's phosphate-buffered saline (PBS pH 7.4, containing 2% antibiotic–antimycotic solution) and collect on a 20 μm nylon filter mesh.
3 Digest the tissue with constant mechanical agitation with 0.5% collagenase (in calcium-, magnesium-free PBS containing 0.5% BSA) at 25 °C for 1 h. (Note: Although the authors do not give a ratio of digestion buffer volume to tissue volume, our experience has shown a ratio of 5:1 is the minimum for effective digestion and that digestion at 37 °C in a rotating water bath is more effective.)
4 Centrifuge the tissue (450 *g*, 4 min) and wash two more times in PBS.
5 Resuspend the pellet in PBS and filter the suspension through a 100 μm nylon mesh.
6 Centrifuge the filtrate and resuspend in growth medium (DMEM, 10% plasma derived serum, 1% retinal growth extract,* 90 μg/ml heparin and 2 mM L-glutamine). (Note: This medium was found to enrich for endothelial cells and was necessary at an early stage to ensure homogeneous endothelial cell populations later.) Then plate on gelatin-coated plates (Bacto gelatin (Difco, Detroit, MI) 1.5% in CMF-PBS) or Primaria

* Retinas were removed from bovine eyes and incubated for 3 h at 25 °C in HBSS (1 retina/1 ml). The mixture was centrifuged (600 *g*) for 5–10 min at 25 °C. The supernatant was passed through a 0.2 μm filter (Nalgene, Rochester, NY)

plates and grow in a humidified incubator at 37 °C at 10% CO_2. Such a preparation from one lung was plated on 6–35 mm dishes and grown in a humidified incubator at 37 °C at 10% CO_2.

7 Re-feed cells after 2 h with plating medium to remove non-adherent cells.
8 Passage the cells at confluence (5–7 days) at a 1:3 split using pancreatin, followed by trypsin/EDTA.
9 Maintain the cells in growth medium and passage every 7–10 days until cultures 'show endothelial cell growth' by phase microscopy. At this point replace the serum by 10% FBS (Hyclone Labs, Logan, UT) and for subsequent passages use trypsin. The cells have been used for up to 16 passages reproducibly. (The authors do not give a medium for freezing but our best success with any cell line is with 10% DMSO and 90% FBS.)

The cells were characterised by the presence of factor VIII related antigen and uptake of DiI-Ac-LDL (both from Biomedical Technologies, Stoughton, MA), angiotensin converting enzyme (ACE; Dr J.J. Lanzillo, New England Medical Center, Boston, MA), by ACE activity (Ventrex Labs, Portland, ME) and by scanning microscopy which revealed numerous surface projections on the cells that were not seen in corresponding large vessel bovine endothelium. (It should be noted that Del Vecchio *et al.* (1992) did not see such projections, which may indicate differences in the cell isolates.) In subsequent work by the same laboratory, cytokeratin 18 (anti-cytokeratin 18; Sigma, St Louis, MO) was shown to be specific for bovine microvessel endothelial cells.

Rabbit pulmonary microvessel endothelial cells

Again, initial isolations of rabbit endothelium from whole lung have been approached by bead perfusion (Ryan *et al.*, 1982). More recently, lung-homogenate isolations have been used but without a selective protocol (Folkman *et al.*, 1979). We (Carley *et al.*, 1990) have consistently found that rabbit microvessel isolation can be routine from whole lung segments by:

1 Biasing the preparation by taking selected peripheral lung segments that are free of mesothelium.
2 Selecting for endothelium by using RDGF-containing medium.
3 Sorting the cells for DiI-Ac-LDL uptake (Carley *et al.*, 1990; Fig. 2.1).

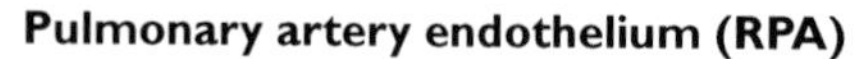

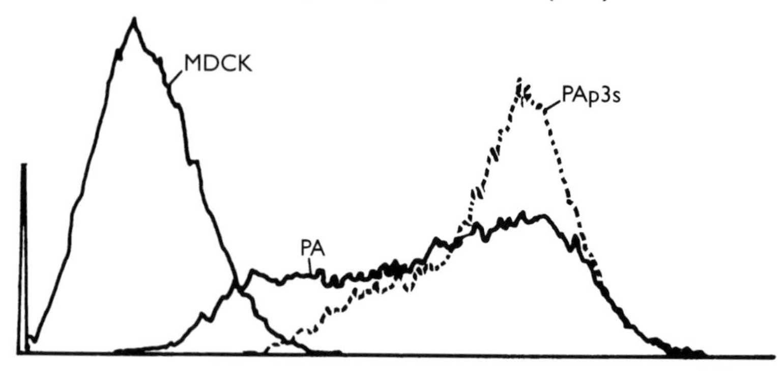

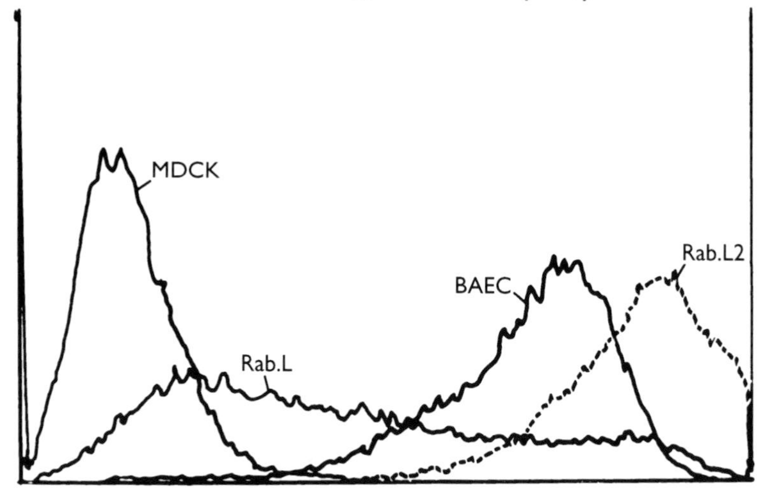

Fig. 2.1. DiI-Ac-LDL uptake profiles of large and small vessel rabbit lung endothelium. Pulmonary artery (PA) or peripheral lung sections (Rab.L) were processed as described under the protocol for rabbit microvessel endothelial cell isolation and growth. Cells were grown out of these preparations (two 75 cm^2 flasks each of PA and Rab.L cells) and sorted for DiI-Ac-LDL uptake by selecting the most fluorescent 1% of the cells. These sorted cells were then grown to confluence and analysed again for DiI-Ac-LDL uptake (Pap3s, third passage after sorting; Rab.L2, second passage after sorting). Madin–Darby canine kidney (MDCK, from ATCC) epithelial cells were used as a negative control and bovine aortic endothelial cells (BAEC) were used as a positive control. The *x*-axis is a reflection of relative cell number (light scatter) and the *y*-axis is log fluorescence.

Protocol for rabbit microvessel endothelial cell isolation and growth

1 Anaesthetise New Zealand white rabbits with an allobarbitol (400 mg/kg) and urethane (100 mg/kg i.v.) solution. Perfuse lungs free of blood cells with Krebs–dextran solution (Carley *et al.*, 1990) for 20 min at 50 ml/min.
2 Remove the lungs from the thorax and immerse in 70% ethanol for 2 min. Mince peripheral lung segments (~1×1×1 cm cubes) within 5 mm of the lung border with a small surgical scissors in DMEM containing 100 U/ml penicillin and 100 μg/ml streptomycin (Note: The pieces need to be minced finely enough to be taken up by a 10 ml pipette and also to ensure they do not float when centrifuged.)
3 Take up the segments with a 10 ml pipette and centrifuge at 800 *g* for 5 min, then resuspend in 0.2% collagenase (Worthington type II) and 0.1% BSA in PBS (we use a 10:1 ratio of enzyme solution to packed tissue volume) and incubate for 30 min in a 50 ml sterile centrifuge tube that is agitated in a 37 °C water bath.
4 After 30 min remove the tube from the bath and draw the partially digested segments up and down with a 5 ml pipette 10 times to help disperse the segments.
5 Incubate the segments under the same conditions for another 30 min and centrifuge at 800 *g* for 10 min.
6 Resuspend the cell pellet in 24 ml Growth Medium [DMEM containing 10% FBS (Hyclone, Logan, UT), 20 μg/ml heparin (Fischer Scientific, Fairlawn, NJ), 5 μl/ml RDGF (made as described in the footnote on p. 11) and 4 mM glutamine (Gibco, Grand Island, NY)]. Place an aliquot of 6 ml into a 25 cm^2 tissue culture flask coated with 1.5% gelatin in PBS (see bovine cell protocol).
7 After 24 h wash the cells with 10 ml medium and re-feed with 6 ml. Then grow these cells for about 1 week until confluent.
8 For FACS purification, incubate one 75% confluent 75 cm^2 flask with 6 ml Growth Medium containing 5 μg/ml DiI-Ac-LDL for 4 h at 37 °C.
9 Trypsinise the cells, resuspend in 1 ml Growth Medium and sort on a FACstar Plus flow cytometer (Becton-Dickinson Immunocytometry Systems, San Jose, CA). An argon-ion laser emitting 200 mW at 488 nm provides the excitation source for the DiI-Ac-LDL, and gate is set such that the top 0.5–1.0% of the cells are sorted. The usual yield from such a sort is 10 000 cells which can then be plated in two 35 mm diameter gelatin coated dishes. (Care must be taken to fill the FACS collection tube to within 1 cm of the top, so the collected cells do not dry while being

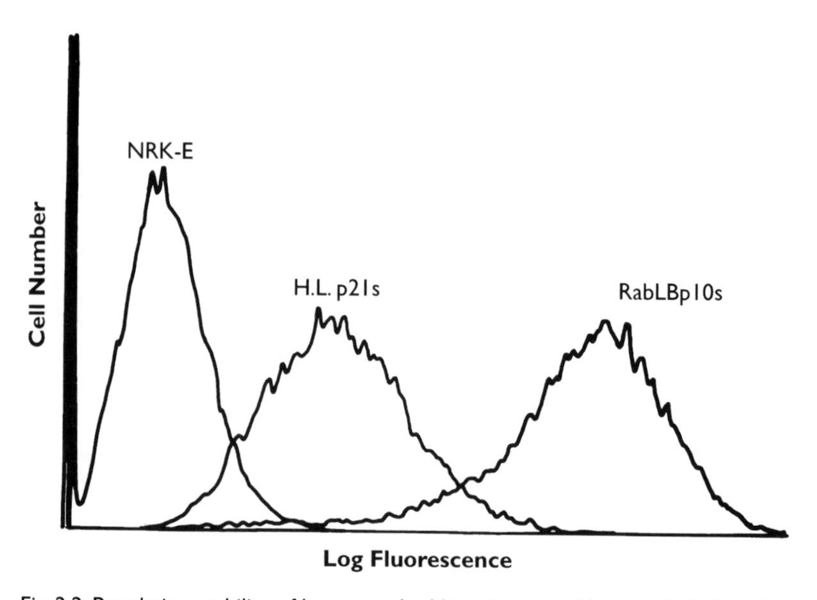

Fig. 2.2. Population stability of human and rabbit microvessel lung endothelium. DiI-Ac-LDL sorted human and rabbit microvessel endothelium were carried to 21 (H.L. p21s) or 10 (RabLBp10s) passages, respectively, after sorting and analysed for DiI-Ac-LDL uptake. Normal rat kidney (NRK-E) epithelial cells were used as a negative control.

sorted. Also, the top of the collection tube should be repeatedly washed to recover all the collected cells.)

Rabbit microvessel (RMV) cells isolated with this protocol can be grown and FACS-analysed repeatedly to ensure phenotypic homogeneity (see Rab.L2 sort in Fig. 2.1) as compared with the initial population (see Rab.L sort in Fig. 2.1). If macrophages are a contaminant they will appear as cells even 1 log more fluorescent than the positive population and will not grow. This sorted population is stable for at least 10 passages (Fig. 2.2).

The same protocol can be applied to sorting rabbit pulmonary artery that has been minced and processed in a similar manner (see Fig. 2.1, RPA) to provide a pulmonary large vessel population for comparative studies. The RPA cells consistently incorporated less DiI-Ac-LDL.

Markers for these cells include: Factor VIII related antigen and DiI-Ac-LDL (both from Biomedical Technologies, Stoughton, MA), angiotensin converting enzyme (ACE; Dr R. Soffer, Cornell University, New York, NY) by ACE activity (Ventrex Labs, Portland, ME) and by FITC-conjugated BSI-B_4 binding (Sigma, St Louis, MO).

Fig. 2.3. Silver staining of rabbit lung endothelial cells. Rabbit lung microvessel endothelial cells (passage 8 after sorting) were plated on gelatin/fibronectin coated glass coverslips (1.5% gelatin, 100 μg/ml human fibronectin; Collaborative Biomedical Products, Bedford, MA). The cells were grown to confluence in growth medium and then for 8 days in medium without RDGF (see footnote on p. 11) and silver stained (See 'Protocol for rabbit microvessel endothelial cell isolation and growth'). Total field width was 680 μm.

In an effort to show that these same rabbit microvessel cells retained the capacity to mimic *in vivo* capillary hydraulic conductivity (L_p) and albumin reflection coefficient ($\sigma_{albumin}$) Powers & Staub (1993) modelled a monolayer in microporous filter chambers. Values for L_p in the presence of growth factor were 37×10^{-7} cm/(s×cm H_2O), while they approached values seen in frog mesentery when the growth factor was removed [6×10^{-7} cm/(s×cm H_2O): Powers & Staub, 1993; M. Powers, personal communication]. These results were mimicked by the ability to silver stain the monolayer of cells being measured (Fig. 2.3) and may, therefore, represent a relatively easy method for defining a marker for continuous capillary endothelial cells *in vitro* in the future. The protocol for the staining is given below.

Protocol for silver staining RMV cells

1 Grow cells to confluence on glass coverslips in medium containing RDGF. Wash the monolayer with RPMI without serum or RDGF and grow the cells for a further 5–7 days.

2 Remove the medium and fix the cells in 0.5% glutaraldehyde and 1% paraformaldehyde in 0.075 M cacodylate buffer, pH 7.4, for 15 min at 25 °C.
3 Remove the fixative and follow by rapid washes of 0.9% NaCl, 5% glucose in water and 0.2% $AgNO_3$ (resident for 5–10 s), 5% glucose in water (10 s) and, finally, fixative again for 3 min.
4 Develop silver halide by a 15 min exposure to light from a 150 W bulb (GE model EKE) in a high-intensity fibre optic illuminator (80% of maximal intensity; model 180, Dolan-Jenner, Woburn, MA).
5 Keep the cells in fixative overnight and then in 50%, 70%, 95% and 100% ethanol. Keep cells in 100% ethanol overnight.
6 Clear cells in toluene overnight and mount in Permount (Fischer Scientific, San Francisco, CA) for photographing.

Human pulmonary microvessel endothelial cells

Human pulmonary microvessel cells have historically been difficult to obtain due to accessibility of healthy tissue and general growth requirements. Notably two laboratories have contributed to finding methods for growing and characterising pulmonary microvessel cells (Carley *et al.*, 1993; Hewett & Murray, 1993). Since the results differ, it is worthwhile examining the differences in isolation techniques. Fortunately, Hewett & Murray (1993) were dogged in their approach and used two techniques: (1) one similar to Chung-Welch *et al.* (1988) and Carley *et al.* (1992); and (2) another that employed UEA-Dynabeads. The two laboratories initial isolation approaches are similar to that of Chung-Welch, *et al.* (1988) in that tissue was isolated free of pleura, minced and digested. But the similarity stops there. Table 2.1 highlights the differences in protocols.

In both laboratories technique 1 yielded cells that incorporated DiI-Ac-LDL (Carley *et al.*, 1992; Hewett & Murray, 1993; and Fig. 2.2 'H.L. p21s'), stained weakly for vWF, immunostained for ACE and bound UEA. Also, these cells did not express EN4, PAL-E, PECAM-1 or E-selectin (after TNF stimulation). The regulation of IL-8 and MCP-1 was characterised in these cells as well (Brown *et al.*, 1994). Carley *et al.* (1992) found that the cells rapidly formed tubes and degraded the extracellular matrix, perhaps due to their ability to produce 200-fold more urokinase plasminogen activator than human umbilical endothelial cells (HUVEC) (Carley *et al.*, 1992; Gerritsen *et al.*, 1993). Hewett & Murray (1993) termed these cells 'UCP', unidentified cell population.

Hewett & Murray (1993) took an alternative approach to isolation by

Table 2.1. *Comparison of human lung microvessel cell isolation techniques*

	Technique 1 (Carley *et al.*, 1992)	Technique 2 (Hewett & Murray, 1993)
Dispersion enzyme(s):	0.1% collagenase/0.2% BSA	2 U/ml dispase+trypsin/EDTA
[time]:	1 h	16 h+1 h
At this point, both protocols then passed the cells through a 100 μm mesh but grew the cells in different medium.		
Growth medium:	RPMI/10% FCS/10% Nu serum, 20 mg/ml heparin, 4 μl/ml RDGF, 2 mM glutamine, Pen-Strep	M199/0.014 M HEPES/0.15% sodium hydrogencarbonate, 2 mM glutamine, 90 μg/ml heparin, 30% FCS, 15 μg/ml endothelial growth supplement
The cells were grown on a coating of:	1.5% gelatin/PBS	5 μg/cm^2 fibronectin
Cell selection:	DiI-Ac-LDL (This step was not used in Hewett's initial isolates)	UEA-Dynabead separation (details in text)

increasing the digestion time and by (repeatedly) selecting the cells that bound UEA. These cells exhibited 'typical endothelial "cobblestone" morphology', expressed vWF, ACE, thrombomodulin, EN4, PAL-E, H4-7/33, PECAM-1 and E-selectin upon TNFα stimulation, and incorporated DiI-Ac-LDL. These cells have many of the same markers as HUVEC and high endothelial venules – both lines of venular origin.

The simple conclusion from these results is that technique 1 enriches for the capillary endothelium and technique 2 for venular and/or arteriolar endothelium. More significantly, the results emphasise that initial isolates may contain a representative population of endothelium from the lung but that molecular selection of the population is immediately necessary to isolate those cells which may either be in a minority in the organ (venules) or may be at a growth rate disadvantage when compared with others (capillaries). In angiogenesis assays it is the capillaries that invade and grow so quickly. Additional molecular marker identification for these two cell lines may fully answer this query.

These results also exemplify the power of the molecular-directed Dynabead approach. Any ligand that binds preferentially to endothelium *in situ* can theoretically be used for isolating microvessel cells without the expense of

FACS equipment (Chung-Welch *et al.*, 1988). Because of this technology Dynabead conjugation and their use in cell isolation are detailed below.

UEA-1 conjugation to Dynabeads

Sterile-filtered *Ulex europaeus* agglutinin-1 (Sigma, St Louis, MO) at 0.2 mg/ml was added to an equal volume of tosyl-activated Dynabeads M450 (Dynal, Merseyside, England) in 0.5 M sodium tetraborate buffer (pH 9.5) and mixed for 24 h on a rotary shaker. The beads were washed four times and incubated overnight with sterile 0.1% BSA in PBS. Such conjugates are active for at least 4 months.

UEA-1 Dynabead usage

The beads should be used at the first cell passage after plating. Cells are passaged in trypsin (Sigma) and 1 mM EDTA. HBSS with 5% FCS was added and the cells pelleted at low speed. Cells (10^5–10^6) were resuspended in the same buffer and incubated with 30 μl (1–2×10^7) of UEA-1 beads for 10 min at 4 °C with occasional agitation. HBSS/BSA buffer was added to a final volume of 10 ml and the suspension was selected in a magnetic particle concentrator (MPC-1, Dynal) for 3 min. Such a washing protocol was repeated 3–5 times with 12 ml volumes and the selected cells plated onto gelatin (0.2%) coated tissue culture flasks. Cells were grown and split at a 1:4 ratio.

Summary

Techniques have been detailed for isolating and growing rat, mouse, bovine, rabbit and human pulmonary microvessel endothelium. Relatively simple, straightforward basic protocols are described which allow the isolation of all lung cells as well as methods for growth enrichment, physical enrichment and molecular marker-enrichment for endothelial cells. In addition, two techniques are described which allow for enrichment of what are thought to be subpopulations of microvessel cells from the lung. Such *in vitro* approaches exemplify the progressive refinement of culture techniques which will aid the physiologist as well as the molecular biologist in eventually extrapolating single cell population behaviour to whole organ physiology.

References

Brown, Z., Gerritsen, M.E., Carley, W.W., Streiter, R., Kunkel, S. & Westwick, J. (1994). Chemokine gene expression and secretion by cytokine-activated human microvascular endothelial cells: differential regulation of monocyte chemoattractant protein-1 and interleukin-8 in response to interferon-γ. *Am. J. Pathol.*, **145**, 913–21.

Carley, W.W., Tanoue, L., Merker, M. & Gillis, C.N (1990). Isolation of rabbit pulmonary microvascular endothelial cells and characterization of their angiotensin converting enzyme activity. *Pulm. Pharmacol.*, **3**, 35–40.

Carley, W.W. , Niedbala, M.J. & Gerritsen, M.E. (1992). Isolation, cultivation and partial characterization of microvascular endothelium derived from human lung. *Am. J. Resp. Cell Mol. Biol.*, **7**, 620–30.

Chung-Welch, N., Shepro, D., Dunham, B. & Hechtman, H.B. (1988). Prostacyclin and prostaglandin E_2 secretions by bovine pulmonary microvessel endothelial cells are altered by changes in culture conditions. *J. Cell Physiol.*, **135**, 224–34.

Davies, P., Smith, B., Maddalo, F., Langleben, D., Tobias, D., Fujiwara, K. & Reid, L. (1987). Characterization of lung pericytes in culture including their growth inhibition by endothelial substrate. *Microvasc. Res.*, **33**, 300–14.

Del Vecchio, P.J., Siflinger-Birnboim, A., Belloni, P.M., Holleran, L., Lum, H. & Malik, A.B. (1992). Culture and characterization of pulmonary microvascular endothelial cells. *In Vitro Cell Dev. Biol.*, **28A**, 711–15.

Folkman, J., Haudenschild, C.C. & Zetter, B.R. (1979). Long-term culture of capillary endothelial cells. *Proc. Natl. Acad. Sci. USA*, **76**, 5217–21.

Gerritsen, M.E., Chen, S.-P., McHugh, M.C., Atkinson, W.J., Kiely, J.-M., Milstone, D.S., Luscinskas, F.W. & Gimbrone, M.J. (1995). Activation-dependent isolation and culture of murine pulmonary microvascular endothelium. *Microcirculation*, **2**, 151–63.

Gerritsen, M.E., Niedbala, M.J., Szczepanski, A. &. Carley, W.W (1993). Cytokine activation of human macro- and microvessel-derived endothelial cells. *Blood Cells*, **19**, 325–42.

Habliston, D.L., Whitacker, C., Hart, M.A., Ryan, U.S. & Ryan, J.W. (1979). Isolation and culture of endothelial cells from the lungs of small animals. *Am. Rev. Respir. Dis.*, **119**, 853–68.

Hewett, P.W. & Murray, J.C. (1993). Human lung microvessel endothelial cells: isolation, culture and characterization. *Microvasc. Res.*, **46**, 89–102.

Magee, J.C., Stone, A.E., Oldham, K.T. & Guice, K.S. (1994). Isolation, culture and characterization of rat lung microvascular endothelial cells *Am. J. Physiol.*, **267**, L433–41.

Powers, M. & Staub, N. (1993). Barrier properties of cultured rabbit lung microvascular endothelial cells. *FASEB J.*, **7**, A898.

Ryan, U.S., White, L., Lopez, M. & Ryan, J. (1982). Use of microcarriers to isolate and culture pulmonary microvascular endothelium *Tissue Cell*, **14**, 597–606.

3

Bone marrow endothelium

John Sweetenham and Lisa Masek

Introduction

Bone marrow is a highly compartmentalised structure in which haemopoiesis occurs within distinct anatomical niches. Endothelial cells of the bone marrow microvasculature are thought to play a central role in regulating cellular traffic between the intravascular and extravascular (haemopoietic) compartments of the bone marrow (Tavassoli, 1979). The volume of this cellular traffic is highly variable, according to demands such as blood loss or infection. Normal haemopoiesis requires intimate contact between marrow stromal cells and haemopoietic progenitor cells (HPCs). Maturing HPCs detach from the stroma, traverse the sinusoidal endothelium and enter the circulation.

Circulating HPCs are capable of selective 'homing' to the bone marrow (Tavassoli & Hardy, 1990). This process is essential to the clinical practice of bone marrow transplantation, in which intravenously transplanted progenitors return to the extravascular compartment of the bone marrow and restore multilineage haemopoiesis. Homing is also thought to occur under normal conditions of haemopoiesis. The homing process is thought to occur in two stages. In the first step, circulating HPCs interact with the luminal surface of sinus-lining endothelium. The cells then migrate through the endothelial layer and enter the extravascular compartment, where they interact with stromal cells, extracellular matrix components, and growth factors.

Interactions between various stromal components of the bone marrow, and HPCs are well documented, and the role of these interactions in progenitor cell homing has been the subject of a large number of studies. By contrast, little is known about the role of microvascular endothelium of the bone marrow in the homing process. This has been due mainly to the lack of reliable cytochemical and histochemical markers for these cells, and difficulties in isolation and culture of the cells for *in vitro* studies. In the last 2 years there has been significant progress in identification, isolation and culture of human bone marrow microvascular endothelium, and *in vitro* adhesion assays to allow dissection of the mechanisms involved in progenitor cell homing have been developed.

Further characterisation of marrow endothelium may have several potential applications. The egress of HPCs from the haemopoietic compartment of the marrow may be partly under the control of the sinus-lining endothelium. This process is fundamental to the body's response to conditions such as blood loss or infection. In addition, in the clinical practice of peripheral blood progenitor cell (PBPC) transplantation, HPCs are 'mobilised' from the marrow into the circulation by a combination of cytotoxic chemotherapy and cytokines such as granulocyte-colony stimulating factor (G-CSF). Further understanding of the adhesion molecules involved in HPC homing may lead to strategies which produce more rapid haematological recovery after high dose chemo or radiotherapy. In addition, study of these cells will allow investigation of their potential role in mechanisms of cancer cell metastasis to the bone marrow.

Characterisation of bone marrow endothelial cells

Histochemical markers

Histochemical characterisation of human marrow endothelium *in situ* has recently been reported from our own laboratory, using routinely processed bone marrow biopsies and freeze-substituted and resin-embedded tissue (Masek *et al.*, 1994). We were able to demonstrate that markers in routine use for the detection of endothelial cells in other tissues can also be used in the bone marrow. The two most widely used markers, antibodies to von Willebrand factor [anti-vWF, anti-factor VIII related antigen (anti-FVIIIR-Ag)] and the plant lectin *Ulex europaeus* agglutinin-1 (UEA-1), produced strong staining of human marrow endothelium. In addition, positive staining was seen with the monoclonal antibody BMA120. This antibody reacts with a partially characterised 200 kDa glycoprotein present in the cytoplasm and on the cell membrane of human endothelial cells from several anatomical sites. Positive staining was also demonstrated with the QBend 10 monoclonal antibody which reacts with an epitope of the CD34 antigen, present on the surface of endothelial cells.

We were also able to demonstrate microheterogeneity of marrow endothelial cells, based on their expression of alkaline phosphatase.

Table 3.1. *Immunocytochemical staining of cultured human bone marrow endothelial cells*

Antibody	Staining[a]
FVIIIRa	++
CD31	+/−
BMA120	+
CD34 (QBend 10)	−
ICAM-1	++
E-selectin	+/−
VCAM-1	++
CD11a	−

Notes:
Cells were stimulated with 10 ng/ml recombinant human tumour necrosis factor α for 4–6 h prior to staining with ICAM-1, E-selectin and VCAM-1.
FVIIIRa, factor VIII related antigen; ICAM-1, intercellular adhesion molecule-1; VCAM-1, vascular cell adhesion molecule-1.
[a] Scoring was as follows: −, negative; +/−, positive staining in some cells; + to +++, varying degrees of reactivity in all cells.

Prominent alkaline phosphatase reactivity was seen in most endothelial cells in the bone marrow. However, whilst enzyme inhibitors such as lysine and levamisole produced complete inhibition of enzyme activity in all vessels, sodium arsenate and L-leucyl-glycyl-glycine inhibited reactivity in arterioles but not in sinus-lining endothelium.

In addition to the above cytochemical markers, Schweitzer *et al.* (1995) have identified two other monoclonal antibodies with apparent specificity for marrow endothelial cells. The BNH9 monoclonal antibody, reactive with H and Y antigens, and the S-Endo-1 antibody, reactive with a 110 kDa surface antigen on human umbilical vein endothelial cells, both produced positive staining of endothelial cells in frozen sections of human bone marrow.

We, and other groups have further characterised these cells cytochemically and by fluorescence-activated cell sorting (FACS) analysis of cultured cells. Table 3.1 shows the results of immunostaining of cultured human marrow endothelial cells, with examples of staining in Fig. 3.1. The cells show strong positive staining for vWF, BMA120 and CD31. We have been unable to demonstrate reproducible positive staining for CD31 in bone marrow

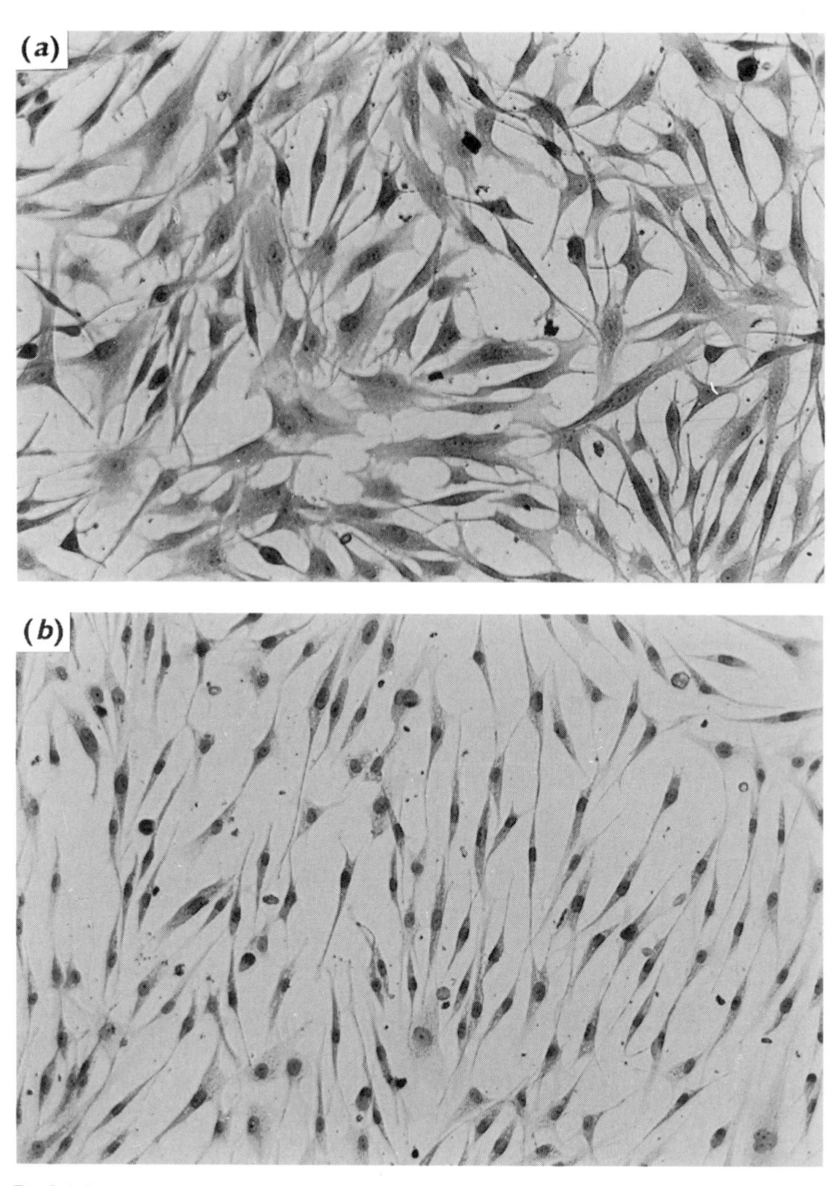

Fig. 3.1. Immunostaining of cultured human bone marrow endothelial cells (HBMEC) with antibodies to vWF (*a*) and CD31 (*b*).

biopsies, although other groups have reported positive staining. We have also shown strong positive staining of these cells for intercellular adhesion molecule-1 (ICAM-1) and vascular cell adhesion molecule-1 (VCAM-1), as well as E-selectin. Staining for all these antigens is increased by activation of the cells with recombinant human tumour necrosis factor α (TNFα). Flow cytometric analysis of the isolated cells also shows a high level of positive staining with vWF and BMA120.

Schweitzer *et al.* (1995) have reported very similar results from immunostaining of sorted and cultured marrow endothelial cells. Over 90% of the isolated cells showed positive staining for vWF, E-selectin, ICAM-1, VCAM-1 and CD31. Rafii *et al.* (1994) have also demonstrated positivity for vWF and CD31 in cultured cells.

Staining for CD34 has been inconsistent. We have shown positive staining for CD34 in paraffin-embedded tissue using the anti-CD34 antibody QBend 10, but have been unable to produce positive staining with this antibody in late passage cultured cells. We have not assessed CD34 staining in early passage cells to date. Using different anti-CD34 antibodies, HPCA-1 and HPCA-2, positive staining of early passage cultured cells has been reported by other groups (Rafii *et al.*,1994; Scweitzer *et al.*, 1995). As with CD34 expression on other endothelial cells, this is lost after one or two passages.

Morphological characteristics

The morphology of cultured human bone marrow endothelial cells (HBMEC) when grown as a monolayer is shown in Fig. 3.2. Using our isolation method (see below), early cultures, at 2–4 weeks are characterised by the presence of two morphologically distinct cell populations. One has a rounded morphology and slow growth rate and the other has a more spindle-shaped morphology, a more rapid growth rate, and tends to become the predominant cell type. Both cell types show positive staining for the markers described earlier. Both cell types have multiple cytoplasmic vesicles (Fig. 3.3). Using immunomagnetic beads and FACS for isolation of cells, Schweitzer *et al.* (1995) report spindle-shaped morphology of HBMEC very similar to that which we observe. Rafii *et al.* (1994) describe spindle-shaped and more typical cobblestone morphology for HBMEC isolated from microvessel explants from human marrow.

Electron microscopy

Electron microscopic features of these cells are shown in Fig. 3.4. The cells have a single rounded nucleus with peripheral chromatin condensation, usually with one nucleolus. The cytoplasm contains numerous membrane-bound vesicles, clusters of free ribosomes and numerous elongated mitochondria. Densely staining Weibel–Palade bodies are present. Similar

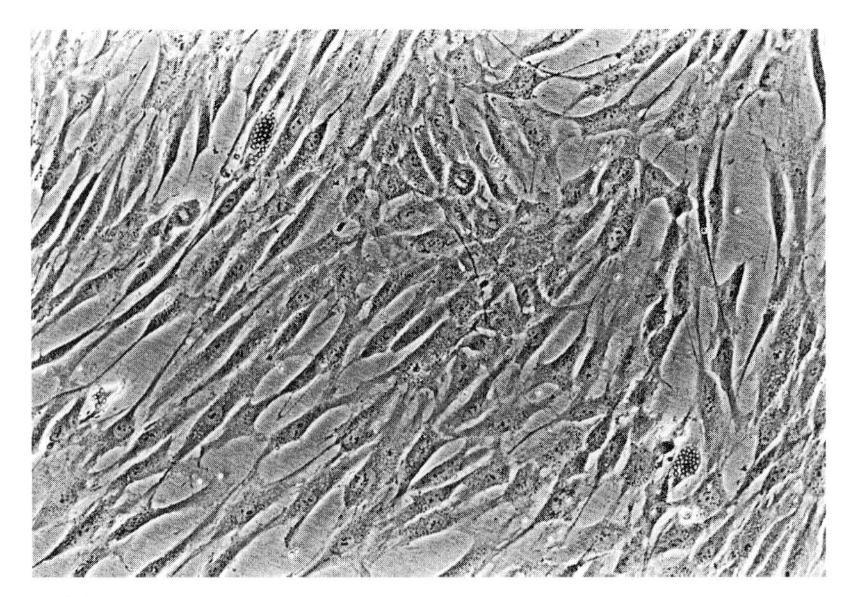

Fig. 3.2. HBMEC cultured on human umbilical vein endothelial cells (HUVEC) extracellular matrix after 2 weeks, showing cells with predominantly spindle-shaped morphology, some still containing beads (original magnification ×50). Reproduced from Masek & Sweetenham (1994).

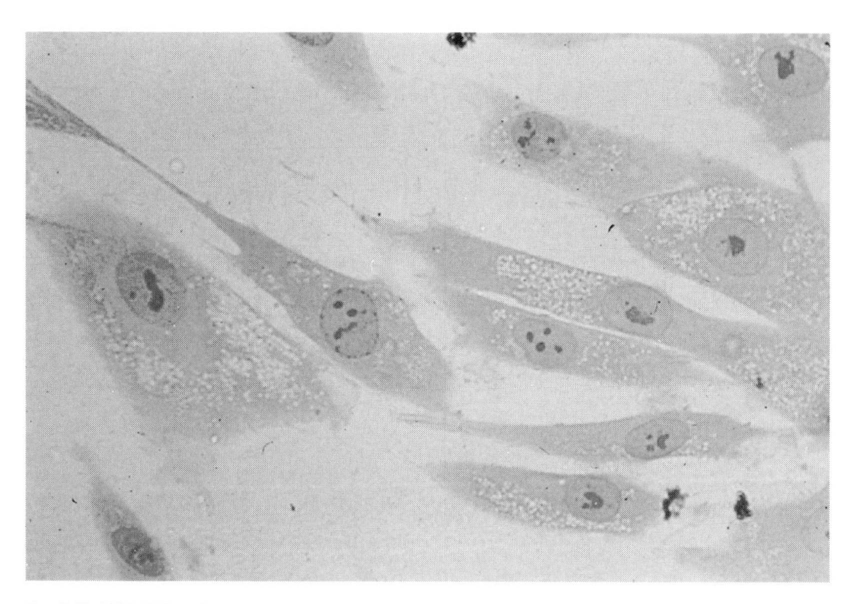

Fig. 3.3. HBMEC cultured on HUVEC extracellular matrix. High-power photomicrograph demonstrating cytoplasmic vesicles.

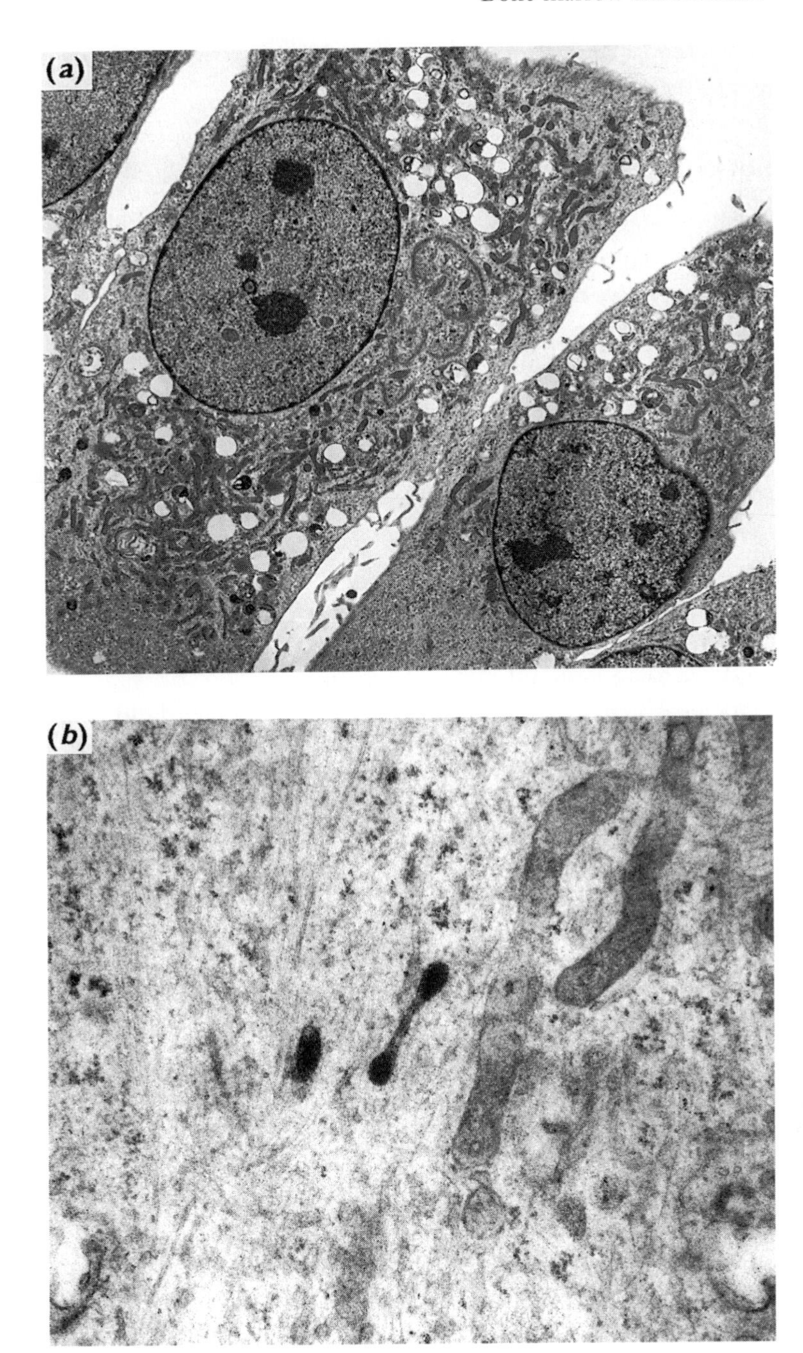

Fig. 3.4. Electron micrograph of cultured HBMEC in contact, grown on fibronectin (*a*) and showing Weibel–Palade body (*b*). Reproduced from Masek & Sweetenham (1994).

electron microscopic appearances have been noted in the two other reports of isolation of these cells.

Source of cells

Adequate cells can be obtained for endothelial cell culture from routine bone marrow aspirates similar to those performed for diagnostic/staging purposes. In our initial studies, we obtained large volumes (40–80 ml) of bone marrow from patients undergoing bone marrow harvests prior to autologous bone marrow transplantation, or from allogeneic bone marrow donors. More recently, we have used routine marrow aspirates, comprising between 3 and 10 ml of particulate bone marrow. Whilst bone marrow from normal healthy donors is probably the ideal source of cells, we have most commonly obtained marrow from patients undergoing staging for malignant lymphoma, in whom the marrow is uninvolved. This has proved to be an adequate source of cells. When allogeneic bone marrow donors are used as a source of cells, sufficient cells for endothelial isolation can be obtained by flushing the filters of Baxter/Fenwal bone marrow collection kits with Hank's Balanced Salt Solution (HBSS).

Other potential sources of human marrow include femoral heads from patients undergoing hip replacement or with hip fractures, and patients undergoing rib resections during thoracic surgery. We have found these to be unsatisfactory sources of marrow for endothelial isolation, and have abandoned their use.

Isolation methods

Chemicals/reagents

Chemicals, unless otherwise stated, are from Sigma Laboratories, St Louis, MO; tissue culture media, unless otherwise stated, are from Life Technologies Ltd, Paisley, UK.

Preservative-free heparin
Dulbecco's Modified Eagle Medium (DMEM)
Hank's Balanced Salt Solution (HBSS)
Red cell lysis buffer (NH_4Cl 155 mM, $NaHCO_3$ 10m M, EDTA 0.1 mM; pH 7.2)
Fetal calf serum (FCS)
Ulex europaeus agglutinin-1 (UEA-1) lectin, 0.2 mg/ml in borate solution pH 9.5
0.1% bovine serum albumin (BSA)

Equipment
Tissue culture plastic (NUNC), unless otherwise stated, is from Life Technologies Ltd.

70 μm mesh strainer (Falcon, Becton-Dickinson, Oxford, UK)
Tosyl-activated M450-Dynabeads (Dynal (UK) Ltd, Wirral, UK)
Magnetic particle concentrator (Dynal)

There are currently three published methods for isolation of endothelial cells from human bone marrow. The method described by Rafii *et al.* (1994) relies on isolation of microvessels from collagenase digestion of bone marrow spicules taken from marrow aspirates. The microvessels are plated onto gelatin coated plastic dishes and the endothelial cells which are grown out from these microvessel explants are further purified by lectin affinity separation using *Ulex europaeus* agglutinin-1 (UEA-1) coated magnetic beads.

In the method described by Schweitzer *et al.* (1995), mononuclear cells are isolated from bone marrow aspirates by density gradient centrifugation on Ficoll–Hypaque. Subsequent endothelial cell isolation was either performed on these preparations, or the cells were initially established in long-term bone marrow culture (LTBMC), and isolations subsequently performed on trypsinised adherent LTBMC stroma. Isolation was performed by magnetic bead separation, using UEA-1 coated beads, or by flow cytometry. For FACS, the cells were enriched by immunomagnetic depletion with anti-CD2, CD14, CD15 and CD19, followed by FACS using the BNH9 or S-Endo1 antibodies described above.

We have described a much simpler technique for obtaining cultures of high purity (Masek & Sweetenham, 1994).

Preparation of UEA-1 coated magnetic beads

Equal volumes of tosyl-activated M450-Dynabeads and unconjugated UEA-1 lectin (0.2 mg/ml in borate solution, pH 9.5) are incubated for 24 h at room temperature on a rotary mixer. Beads are collected using a magnetic particle concentrator (MPC) and the supernatant discarded. Beads are then washed four times for 15 min, and once more overnight at 4 °C in 5 ml 0.01 M phosphate-buffered saline (PBS)/0.1% bovine serum albumin (BSA) on a rotary mixer. The beads are then collected again using the MPC, resuspended in HBSS+5% FCS at a concentration of 4×10^8 beads/ml and stored at 4 °C.

Bone marrow aspirates, collected as described above, are collected into sterile universal tubes containing heparin (15–20 U/ml) and diluted 1:1 by volume with DMEM. If not used on the day of collection the marrow can be stored in DMEM/heparin overnight at 4 °C. Marrow is then filtered through a 70 μm mesh strainer, washed with 1× HBSS and centrifuged at 200 *g* for 5 min at room temperature. The cells are then subjected to red cell lysis using ammonium chloride lysis buffer for 5–10 min. Cells are centrifuged at 200 *g* for 5 min and resuspended in 5% FCS/HBSS for cell counting.

Cell separation

UEA-1 coated Dynabeads are added to the cells at a bead:target cell ratio of 2:1, assuming endothelial cells to comprise 1–2% of the total cell count. (This ratio is much lower than that described by others, but in our experience higher bead:cell ratios result in very poor attachment of cells to matrix when subsequently plated.) The cells are then incubated on a rotary mixer for 5 min at room temperature. Cells are separated on a MPC for 3 min and non-bound cells removed. Attached cells are washed three times by resuspension in ice-cold HBSS/5% FCS, followed by gentle mixing for 1 min and a further 1 min separation using the MPC. Indirect immunofluorescent staining for vWF can be performed at this stage to confirm the endothelial identity of the cells.

Establishment and maintenance of cultures

Chemicals/reagents

Endothelial serum-free plating medium
Fetal calf serum
Gelatin
Human plasma fibronectin
Bovine corneal endothelial extracellular matrix coated tissue culture plastic
(Biological Industries Ltd, Glasgow, UK)
Endothelial serum-free growth medium
Preservative-free heparin (50 μg/ml)
Penicillin–streptomycin solution (100 IU/ml–100 μg/ml)

According to our isolation and culture method, cells isolated as described above are resuspended in endothelial serum-free plating medium, supplemented with 5% FCS and penicillin–streptomycin solution, and plated onto 12.5 cm^2 (Falcon) or 25 cm^2 tissue culture flasks coated either with 0.1% gelatin, human plasma fibronectin (50 μg/ml), extracellular matrix derived from human umbilical vein endothelial cells (HUVEC: see below), or extracellular matrix from bovine corneal endothelial cells. Cells are cultured in this medium for 24 h at 37 °C in 5% CO_2 in air, after which time the medium is changed to endothelial serum-free growth medium with 5% FCS, 50 μg/ml endothelial cell growth supplement, 50 μg/ml heparin and penicillin–streptomycin solution.

HUVEC isolation and culture

Chemicals/reagents
Type I collagenase
Medium 199 (M199)
Fetal calf serum
Glutamine
Penicillin–streptomycin solution (100 IU/ml–100 μg/ml)
Fungizone solution (5 μg/ml)
Gelatin 0.1%

HUVEC are isolated and cultured according to the method described by Jaffe *et al.* (1973). Umbilical cord is wiped with sterile gauze soaked in 70% ethanol and the ends cut to remove clamp marks. The vein is then gently cannulated with a blunt-ended plastic cannula at each end and flushed with sterile 0.15 M saline to remove remaining blood. Next 0.1% type I collagenase in M199 is introduced into the vein and the cord incubated at 37 °C for 10 min in a saline water bath. Following this, the cord is massaged gently and the released cells are flushed with M199 into a tube containing medium.

The cells are then centrifuged at 200 *g* for 5 min, resuspended in M199 with 20% FCS, 2 mM glutamine, penicillin–streptomycin solution and Fungizone, and cultured in 25 cm^2 tissue culture flasks coated with 0.1% gelatin at 37 °C in 5% CO_2. Cells are subsequently passaged using 1×trypsin/EDTA solution for endothelial cell cultures (Sigma).

Preparation of HUVEC extracellular matrix coated plates

Chemicals/reagents
0.5% Triton X-100 in PBS
Penicillin–streptomycin solution (100 IU/ml–100 μg/ml)

HUVEC are grown to confluence as described above. The identity of the cells can be confirmed at this stage by immunostaining for vWF. Cell monolayers are washed with sterile PBS and incubated with sterile 0.5% Triton X-100/PBS (v/v) for at least 30 min. Three washes are performed with sterile PBS, ensuring that the flasks are tapped and pipetted vigorously to remove all cells, leaving intact extracellular matrix (ECM). The remaining ECM is then rinsed in 10×penicillin–streptomycin solution for 1–2 min. This is discarded, and the coated flasks can then be stored at 4 °C until required.

Long-term culture and passage

Chemicals/reagents
Trypsin/EDTA for endothelial cells

The growth rate of these cells in culture, isolated according to our method, is highly variable, and apparently dependent upon the initial cell density. In general, when grown on HUVEC ECM, the cells attach and the cultures become established more rapidly than with fibronectin or gelatin. With all three substrates the cells reach confluence at between 4 and 6 weeks. They are passaged using trypsin/EDTA for endothelial cells. Split ratios of 1:3 to 1:5 can be used. To date, we have passaged cells up to eight times, after which they appear to lose viability. Most of our *in vitro* studies of adhesion with these cells are carried out on early passage (1 to 5) cells.

Cryopreservation

Freezing

Cells are removed from tissue culture flasks using 1×trypsin/EDTA solution for endothelial cell cultures and centrifuged gently. The supernatant is removed and the pellet resuspended in 0.9 ml of endothelial serum-free medium (growth medium) per 25 cm^2 flask, containing at least 10% FCS. Aliquoted cells are placed in cryotubes and left on wet ice for 10 min. Dimethylsulphoxide is added to a final concentration of 10% and mixed gently. The cells are frozen according to standard protocols and stored in liquid nitrogen.

Thawing

Cells are removed from storage and thawed rapidly in a 37 °C water bath. One millilitre of cell suspension is added to 9 ml of endothelial serum-free medium (growth medium) containing 5% FCS, mixed gently and centrifuged. The pellet is resuspended in endothelial serum-free medium (growth medium) containing 5% FCS, endothelial cell growth supplement, heparin, penicillin and streptomycin as described above, and plated as for primary cultures.

Summary

The method described above allows the culture of human bone marrow endothelial cells of high purity, suitable for *in vitro* study. Studies of adhesion of haemopoietic cell lines to these cells have been developed using a chromium^{-51} labelling assay (Masek *et al.*, 1995). This should provide a useful system for dissecting the various adhesive interactions involved in haemopoietic stem cell homing.

Acknowledgements

We are grateful to Dr Kwee Yong, Department of Haematology, Royal Free Hospital, London, for her assistance with flow cytometry. This work was supported by a grant from the Leukaemia Research Fund.

References

Jaffe, E.A., Nachman, R.L., Becker, C.G. & Minick, C.R. (1973). Culture of human endothelial cells derived from umbilical veins: identification by morphologic and immunologic criteria. *J. Clin. Invest.*, **52**, 2745–56.

Masek, L.C. & Sweetenham, J.W. (1994). Isolation and culture of endothelial cells from human bone marrow. *Br. J. Haematol.*, **88**, 855–65.

Masek, L.C., Sweetenham, J.W., Whitehouse, J.M.A. & Schumacher, U. (1994). Immuno-, lectin-, and enzyme-histochemical characterization of human bone marrow endothelium. *Exp. Hematol.*, **22**, 1203–9.

Masek, L.C, Turner, M.L., Anthony, R.S., Parker, A.C. & Sweetenham, J.W. (1995). Adhesion of human haemopoietic cell lines to human bone marrow endothelial cells (HBMECs) and extracellular matrix (ECM) components: an *in vitro* assay to investigate mechanisms of stem cell homing. *Bone Marrow Transplant.*, **15** (Suppl 2), S38.

Rafii, S., Shapiro, F., Rimarachin, J., Nachman, R.L., Ferris, B., Weksler, B., Moore, M.A.S. & Asch, A. (1994). Isolation and characterization of human bone marrow microvascular endothelial cells: hematopoietic progenitor cell adhesion. *Blood*, **84**, 10–19.

Schweitzer, C.M., van der Schoot, C.E., Drager, A.M., van der Walk, P., Zevenbergen, A., Hooibrink, B., Westra, A.H. & Langenhuijsen, M.M.A.C.

(1995). Isolation and culture of human bone marrow endothelial cells. *Exp. Hematol.*, **23**, 41–8.

Tavassoli, M. (1979). The marrow–blood barrier. *Br. J. Haematol.*, **41**, 297–302.

Tavassoli, M. & Hardy, C.L. (1990). Molecular basis of homing of intravenously transplanted stem cells to the marrow. *Blood*, **76**, 1059–70.

4

Endothelium of the brain

Charlotte Schulze

Introduction

Brain microvascular endothelial cells have successfully been maintained in culture for some time by many groups, but it has proved to be very difficult to establish a true *in vitro* model of the blood–brain barrier (BBB). Although a few promising protocols have been published fairly recently, reproducibility appears to be problematic as judged by the scarcity of studies in which these protocols have been adopted.

The present chapter does not aim at providing an overview on brain endothelial cells in culture. For this the reader is referred to reviews by Rubin (1991), Joó (1992) and Laterra & Goldstein (1992, 1993). Nor is it the intention to present a new protocol for the isolation and growth of these cells in culture. One protocol is given at the end of the chapter, otherwise the reader will be referred to the literature. Instead the focus will be on specific problems associated with the pursuit of establishing an *in vitro* model of the BBB, mainly on aspects such as differentiation of endothelial cells in culture and on evaluation and characterisation of such culture systems. It is meant to be a guide for those who plan to grow brain microvascular endothelial cells in culture.

Brain versus peripheral endothelial cells

Paul Ehrlich observed in 1885 that dye injected into the vasculature of animals readily penetrated all tissues except the brain. About 80 years later it was shown in similar tracer experiments that the structural basis for this 'blood–brain barrier' is the endothelial cells which constitute the microvascular walls in the brain. In these ultrastructural studies it was shown that an electron-dense marker was prevented from extravasation by a combination of impermeable interendothelial junctions and the apparent absence of any transcellular route. Research in the past three decades has revealed that brain endothelial cells are joined to each other by a belt of complex tight junctions. Thin-section images reveal that neighbouring endothelial cells form numerous close contacts within the interendothelial cleft in which the apposing membranes appear to fuse. In freeze-fracture images these tight junctional points of contact appear as a continuous, highly anastomosing network of strands.

Physiological measurements have shown that the transendothelial electrical resistance (TER) across microvascular brain endothelial cells *in vivo* is $>1000\ \Omega\ cm^2$. This reflects a high degree of junctional tightness similar to that found in barrier epithelial cells. In contrast, peripheral microvascular endothelial cells are joined by tight junctions which are far less complex and frequently exhibit discontinuities. They may also have fenestrations, are highly permeable to macromolecules and their TER is typically well below $100\ \Omega\ cm^2$.

A number of features guarantee that the brain interstitium is maintained at a strictly controlled state. Unlike peripheral endothelial cells, brain endothelial cells have a large number of mitochondria and few intracellular vesicles, reflecting a high level of metabolic activity and presumably a low rate of vesicular transport, respectively. Transport of hydrophilic molecules across the BBB is restricted to specific carrier systems, such as the glucose transporter type 1, several amino acid transporters and carriers for amines, nucleosides and purines. Ion homeostasis of the brain interstitial fluid, too, is determined by specific transport mechanisms at the level of the BBB. Fluid-phase endocytosis, i.e. non-specific uptake of substances from the blood, is believed to be very low. In addition, BBB endothelial cells possess drug biotransformation abilities which further increase their protective capacity.

Brain microvascular endothelial cells also differ from those in the periphery in that they have a relatively high pericytic coverage which has been reported to be 50–60% in primates. The corresponding value for muscle tissue is in the order of 10–15%. The functional significance of this difference is not known. Pericytes and endothelial cells share a common basement membrane. It has emerged that pericytes are a heterogeneous population of cells, but their precise functions in most microvascular beds have not yet been elucidated.

A unique feature of BBB microvessels is their close association with astrocytes. Astrocytes have long cytoplasmic cell processes which extend towards the microvessels. These astrocytic footprocesses form a continuous sheath around all BBB vessels (Fig. 4.1).

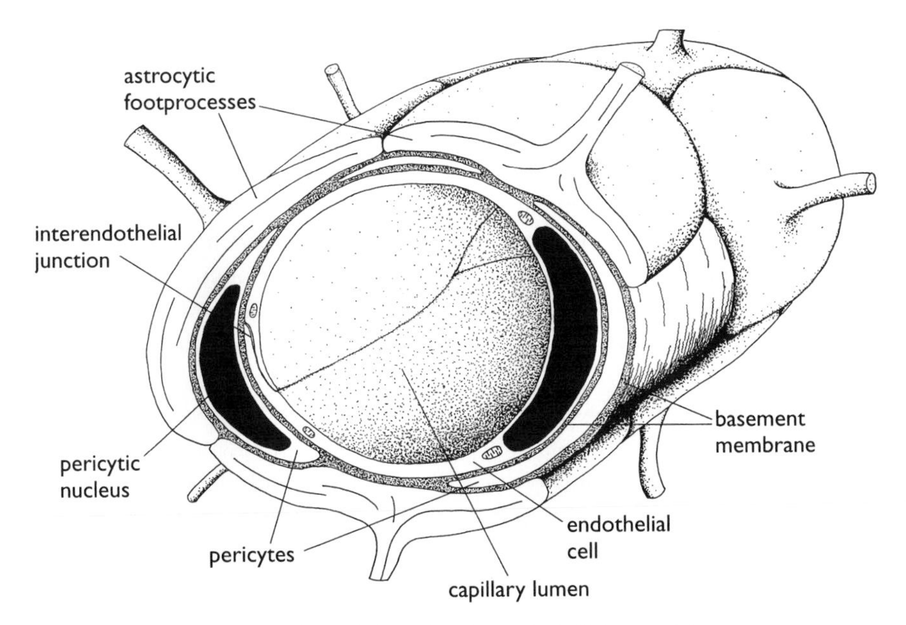

Fig. 4.1. Spatial association of blood–brain barrier endothelial cells with pericytes and astrocytes.

Two classic studies have shed some light on what determines brain endothelial cell differentiation. Stewart & Wiley (1981) have shown that brain endothelial cells which are made to vascularise peripheral tissue lose some of their BBB characteristics; for example, they become leaky to cationic dyes. In contrast, when peripheral endothelial cells vascularise brain tissue they become less permeable. These experiments suggest that the brain endothelial phenotype is not an intrinsic property of these cells but is determined by the surrounding tissue. Janzer & Raff (1987) demonstrated that astrocytes are capable of inducing some BBB characteristics in endothelial cells of non-neural origin, and later evidence largely obtained from cell culture experiments has further supported the notion that astrocytes are capable of inducing BBB differentiation in brain endothelial cells. However, other observations such as the high TER of pial microvessels which are devoid of astrocyte contact or the leakiness of glial cell associated microvessels in the circumventricular organs suggest that other factors are involved. Thus, the process of brain endothelial cell differentiation is likely to be more complex.

Brain endothelial cells in culture

Because of severe experimental limitations on studying the BBB *in vivo*, it has become desirable to grow these cells in culture. The development of an *in vitro* BBB model has happened in essentially three stages. Firstly, viable microvascular fragments were enriched from brain tissue. Secondly, *in vitro* growth conditions were established which allowed attachment and proliferation of brain endothelial cells derived from isolated microvascular fragments. And, thirdly, attempts were made to adapt growth conditions so as to achieve BBB-specific differentiation of these endothelial cells in culture.

Brain microvessels were first successfully enriched in the mid 1970s (reviewed by Takakura *et al.*, 1991; Laterra & Goldstein, 1992, 1993). The softness and homogeneity of brain tissue combined with the relative rigidity of microvessels that are surrounded by a basement membrane, facilitated the isolation of microvessel fragments by the use of simple, mechanical means. Endothelial cells within these fragments were found to be viable and metabolically active for up to a few hours *in vitro*. They were used for physiological, biochemical and transport studies and a considerable amount of insight into the functioning of the BBB was gained from these experiments (Laterra & Goldstein, 1992, 1993). However, this approach was limited by the finite energy supply of the endothelial cells in these preparations and by contamination with other cell types.

Around 1980 the first reports on cultured brain endothelial cells were published (reviewed in Joó, 1992; Laterra & Goldstein, 1992, 1993). Since then many groups have been successful in culturing these cells from a range of species including cow (see references in this article), pig, rat, mouse and human. The isolation procedures employed are variable. For example, brain tissue, usually grey matter, is collected and minced. Further purification steps include combinations of tissue homogenisation, nylon mesh filtration, various centrifugation techniques, enzymatic digestion and trituration. Whereas some procedures include no enzyme digest at all or only a very brief one, others have two digestion steps which can last up to 24 h. The enzymes employed are diverse, as are the concentrations used. Growth conditions such as growth medium, type and concentration of serum, growth supplements or growth factors, the presence of medium conditioned by other cell types and growth substrate also vary greatly, not only between species but also within a single species. The ratio between the amount of starting material and yield, too, is very variable. Whereas some groups use primary cells for their experiments others use later passages or isolate clones. A range of special treatments are used to eliminate contaminating cells, including complement killing of pericytes and selective passaging of cells.

In summary, a multitude of procedures for the isolation and culture of brain endothelial cells from a range of species have been published. However, considerable differences in procedure, possible inherent differences between species and frequently poor characterisation of the resulting endothelial cells render it almost impossible to evaluate the achievements of many groups. A tested protocol is now given.

Reagents, chemicals and equipment

150 μm nylon mesh (Plastok Associates Ltd, UK)
Basic fibroblast growth factor (bFGF; Boehringer Mannheim UK Ltd)
Collagenase (Worthington Biochemical Corp. USA)
CPT–cAMP (Sigma, UK)
Glutamate (Life Technologies Ltd, UK)
Heparin (Sigma, UK)
Human fibronectin (Blood Products Ltd, UK)
Minimum essential medium (Life Technologies Ltd, UK)
Penicillin–streptomycin (Life Technologies Ltd, UK)
Plasma-derived horse serum (Hyclone Laboratories, USA)
RO20-1724 (Calbiochem, UK)
Transwell filters (0.4 μm pore size, polycarbonate, Costar UK Ltd)
Trypsin (Worthington Biochemical Corp.)
Trypsin/EDTA (Life Technologies Ltd, UK)

Protocol for the isolation of bovine brain endothelial cells (after Rubin *et al.*, 1991)

1 Obtain bovine brains fresh from a slaughterhouse and clear of meninges.
2 Collect cortical grey matter in cold L-15 medium and prepare capillary fragments by homogenisation and filtration through a 155 μm nylon mesh.
3 Digest capillary fragments in 0.2% collagenase and 0.04% trypsin for 60 min at 37 °C and then plate on tissue culture flasks in a medium consisting of 50% astrocyte conditioned medium (see below) and 50% Minimum Essential Medium (MEM) supplemented with 10% plasma-derived horse serum. Medium also contains 10 ng/ml FGF and 123 mg/ml heparin. Flasks are pre-coated as described below.

Preparation of astrocyte conditioned medium (after Lillien *et al.*, 1988)

Astrocytes are prepared from a 1-day-old rat cortex as follows:

1 One week after initial mechanical dissociation and seeding, shake the flasks overnight to enrich in type 1 astrocytes.
2 Remove astrocytes with trypsin and seed onto poly-D-lysine coated plates.
3 Treat the cells with cytosine arabinoside (10^{-5} M) and maintain in serum-free medium to limit the growth of rapidly dividing contaminating cells.
4 Prepare conditioned medium following 3–4 weeks of culture by feeding with fresh MEM/10% fetal calf serum (FCS) for 48 h followed by collection, sterile filtration and storage at −20 °C before use.

Procedure for the coating of endothelial culture flasks

1 Coat normal tissue culture plasticware overnight with a solution of rat-tail collagen (100 mg/ml in 1 mM acetic acid).
2 The next day rinse plates with phosphate-buffered saline (PBS) and then coat with human fibronectin (50 mg/ml in PBS) for 2 h.
3 After rinsing three times with PBS the plates are ready for use.

Applications for brain endothelial cell cultures

The availability of brain endothelial culture systems has initiated numerous studies during recent years. In fact, much has recently been learnt about the BBB using this *in vitro* approach.

A large proportion of studies have focused on the characteristic permeability properties of the brain endothelium. These include features such as specific transporters/carriers, receptor-mediated endocytosis, fluid phase endocytosis, ion channels, drug transport and junctional permeability. The recently discovered endothelial cell detoxification properties, involving proteins such as P-glycoprotein, have also attracted considerable attention. A number of groups have studied the influence of other cell types, mostly astrocytes, on BBB endothelial cells. Furthermore, cell culture systems have allowed the study of the effect of cytokines, hormones and other substances on the BBB. The expression of surface markers, whether BBB-specific ones such as γ-glutamyltranspeptidase, or those of a more general nature such as lectin binding sites, have also been the subject of research. Attempts have been made to set up *in vitro* disease models involving the BBB, in the study of, for example, bacterial meningitis. Cell biologists have used brain endothelial cell cultures for comparative studies with other cellular barriers such as the blood–retinal barrier or epithelial barriers. They have also been interested in elucidating signal transduction pathways, for example those involved in the regulation of junctional permeability. One group has tried to identify new BBB-specific genes by subtractive cloning. Some effort has been made to establish stable cell lines that retain BBB characteristics.

This is only a small selection of studies in which brain endothelial cell cultures have been used, illustrating the range of applications for this technique. It is clear that the results obtained in this way are dependent upon the specific properties of the culture system employed. In the majority of cases brain endothelial cells in culture retain typical endothelial cell characteristics, such as cobblestone morphology and immunoreactivity for von Willebrand factor, for a number of passages. In contrast, the expression of BBB-specific markers is usually rapidly downregulated *in vitro*, suggesting that brain endothelial cells tend to de-differentiate and become more like peripheral endothelial cells. Where brain endothelial cells were allowed to form monolayers on semipermeable supports these monolayers were found to be far leakier than their *in vivo* counterparts. Wolburg and coworkers (1994) have recently demonstrated by quantitative freeze-fracture electron microscopy that tight junction complexity of brain endothelial cells is dramatically downregulated within a few hours in culture. Furthermore, the characteristic association of tight junction particles with the P-face is gradually lost. They argue that these morphological changes correlate with increased leakiness of brain endothelial cells in culture. It is obvious, therefore, that results obtained from permeability studies done by use of such a system cannot readily be extrapolated to the *in vivo* situation.

A number of groups have attempted to address these questions by undertaking comparative *in vivo* and *in vitro* studies (Partdridge *et al.*, 1990; Dehouck *et al.*, 1992; Chesné *et al.*, 1993). In some studies a high correlation between *in vivo* and *in vitro* permeability properties was found, whereas in others discrepancies emerged. This approach is often complicated by the fact that *in vivo* information is difficult to obtain or that direct comparisons are methodologically problematic. It is furthermore likely that even a 'good' *in vitro* model of the BBB will differ in some aspects from the *in vivo* situation due to intrinsic limitations of this approach.

Evaluation of *in vitro* models of the blood–brain barrier

All these considerations clearly necessitate a careful characterisation of any *in vitro* model of the BBB before deductions can be attempted. Greenwood (1991) has summarised four main approaches which have been used to evaluate the *in vitro* BBB. Here they will be discussed in a slightly modified form.

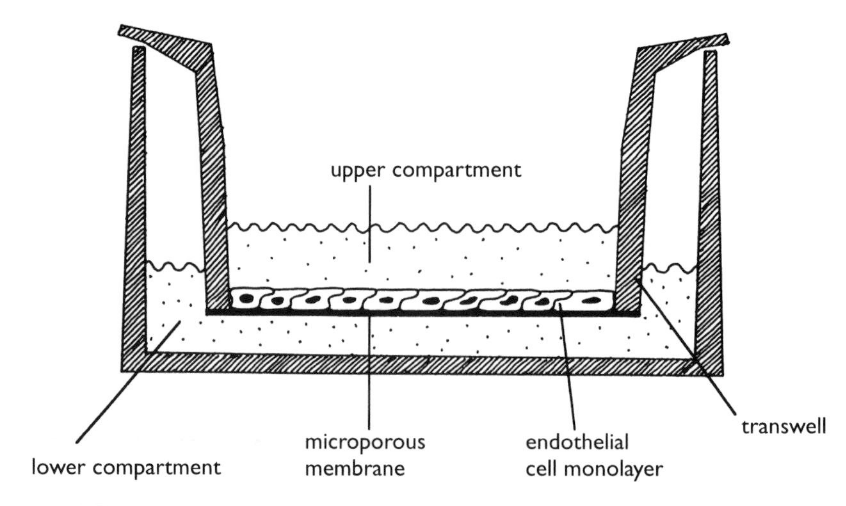

Fig. 4.2. Two-chamber arrangement for culturing brain endothelial cells.

Transendothelial electrical resistance

The electrical resistance measured across a monolayer of cells which are connected to each other by tight junctions is a sensitive indicator for the paracellular permeability to ions as long as transcellular routes are not present. Brain endothelial cells can be grown on permeable filters which separate an upper chamber from a lower chamber (Fig. 4.2). This two-chamber arrangement thus mimics the *in vivo* situation in which a luminal compartment (blood) is separated from an abluminal compartment (brain parenchyma) by a confluent monolayer of endothelial cells. The aim is to achieve electrical resistance across this monolayer of $>1000\ \Omega\ cm^2$ such as has been measured *in vivo*. For this purpose the monolayer needs to be fully confluent and devoid of gaps which may, for example, be formed by contaminating cells that interfere with the monolayer. However, contaminating cells which are growing underneath the monolayer may not directly influence TER. Transcellular gaps and ion channels may also affect TER and care needs to be taken to rule out this possibility. Since most permeable filters are not translucent, direct light microscopical observation of the cells is not possible. However, the monolayers can be fixed and stained by indirect immunofluorescence which does not require transmonolayer illumination.

Transendothelial permeability to non-transported molecules

Monolayers of brain endothelial cells grown on filters can also be used to measure the vectorial transfer of small inert molecules such as sucrose (342

Da) or inulin (approx. 5000 Da) for which no carrier mechanism exists. Such molecules are normally prevented from traversing the BBB. A good *in vitro* model of the BBB should therefore show little or no flux to such molecules.

Tight junction complexity

A number of groups have characterised their brain endothelial cell cultures by freeze-fracture electron microscopy. The images of complex tight junctions were taken as indicators for the presence of a functional barrier. However, freeze-fracture results have to be interpreted with caution. They reflect only very small samples and tight junctions may well be far less complex or even discontinuous in other areas. For these reasons it is important to design a quantitative approach (Wolburg *et al.*, 1994). Furthermore, the degree of complexity may not directly correlate with the degree of tightness. It has also emerged that in the case of brain endothelial cells it is important to analyse both E- and P-faces to get a true picture of the junctional organisation (Wolburg *et al.*, 1994).

Characteristic markers

The characterisation of brain endothelial cells in culture has frequently been rather poor. Due to the availability of an increasing number of antibodies and lectins this task has recently become much easier.

Endothelial cell markers Brain endothelial cells can be screened by immunohistochemistry for the expression of general endothelial markers such as von Willebrand factor. Many companies sell endothelial cell specific antibodies which are often species specific, including the rat-specific anti-RECA-1 antibody. Lectin histochemistry has also been used to characterise endothelial cells.

Blood–brain barrier markers Markers which are specific for the BBB include glucose transporter 1, γ-glutamyltranspeptidase and transferrin receptor. For literature on BBB specific markers the reader is referred to the comprehensive review by Dermietzel & Krause (1991). Some BBB specific antibodies which recognise as yet unidentified epitopes are also commercially available.

Junctional markers Brain endothelial cells are joined to each other by a continuous belt of tight and adherens junctions. Therefore the use of antibodies against molecular components of these junctions provides a good way to

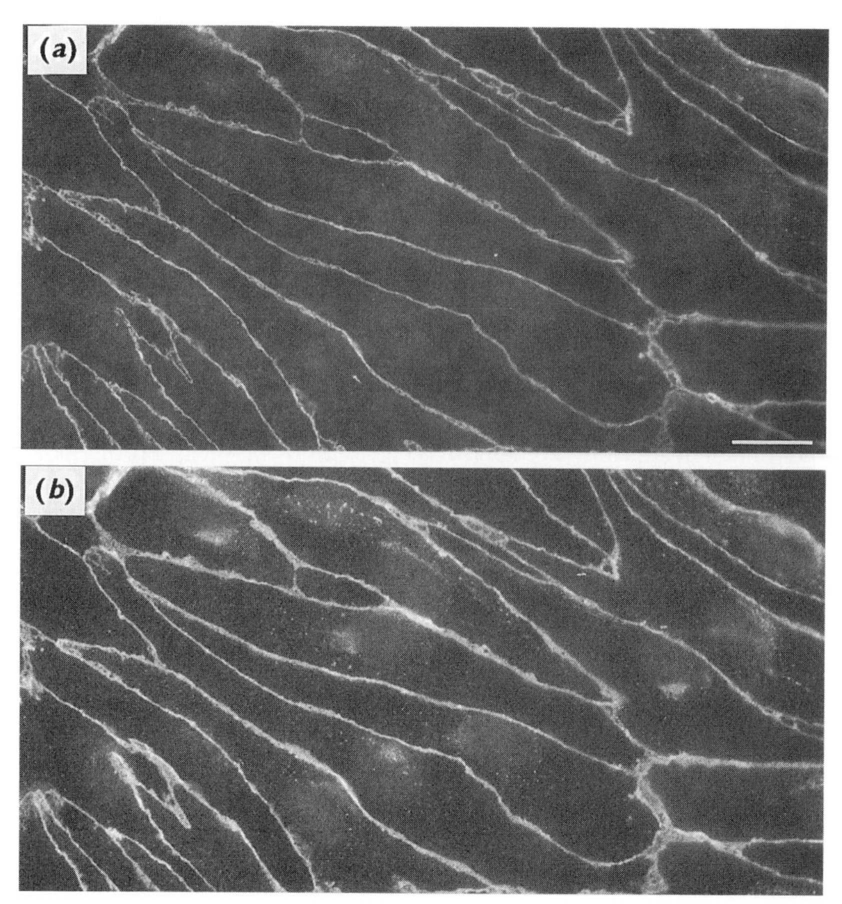

Fig. 4.3. Brain endothelial monolayer double-labelled with the tight junction marker ZO-1 (*a*) and adherens junction marker β-catenin (*b*). Scale bar represents 20 μm.

assess the integrity of the monolayer *in vitro* (Fig. 4.3). Examples are ZO-1 and ZO-2 for tight junctions and members of the catenin family for adherens junctions.

Characterisation of contaminating cells For the evaluation of an *in vitro* BBB model it is also important to assess the presence and identity of contaminating cells. A simple way to reveal the presence of non-endothelial cells is by staining the cultures with phalloidin conjugated to a fluorescent marker. Phalloidin recognises filamentous actin. Well-differentiated brain endothelial cell monolayers are characterised by enriched levels of actin filaments at cell borders with very low levels in other areas. The presence of contaminating cells as revealed by their actin cytoskeleton is well visible above this

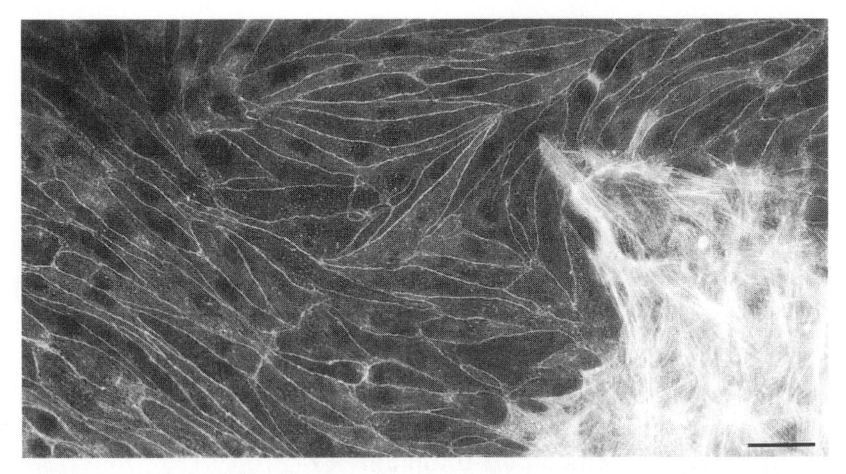

Fig. 4.4. Immunohistochemical detection of actin filaments in a brain endothelial cell monolayer. Pig brain endothelial cells were prepared according to a modification of Rubin *et al.* (1991). A high-resistance monolayer growing on a permeable filter has been fixed and stained with phalloidin coupled to a fluorescent marker. Endothelial cells are characterised by a strong cortical filamentous actin signal. Contaminating cells are visualised by virtue of their actin cytoskeleton. Scale bar represents 50 μm.

background (Fig. 4.4). To identify contaminating cells, cell-specific markers for potential contaminating cells such as smooth muscle actin for pericytes and glial fibrillary acidic protein or S100 for astrocytes can be used.

In order fully to evaluate an *in vitro* model of the BBB no single one of the above-listed markers and properties will be sufficient, and experimenters should strive to combine as many as possible. However, if a monolayer of cultured brain endothelial cells exhibits high TER, this indicates the presence of a confluent, highly impermeable monolayer. Determination of TER is therefore undoubtedly the most sensitive characteristic of all. Additional immunohistochemical characterisation of such monolayers can provide information about the expression of general endothelial cell markers, specific BBB markers and the presence of contaminating cells.

Potential problems

There are a number of potential reasons for failure to generate high TER. The monolayer could simply be subconfluent. A pure population of endothelial cells may not have divided often enough to cover the entire surface of the filter or the presence of contaminating cells may have prevented the endothelial cells from forming a continuous monolayer. Alternatively the monolayer may be confluent but the endothelial cells may have lost some of their BBB characteristics and become leaky. This could indicate that the culture conditions are lacking essential features that allow the endothelial cells to maintain or regain BBB differentiation after a proliferative phase *in vitro*. It is also conceivable that contamination of the culture with endothelial cells derived from large vessels can result in an overall low level of TER, since a considerable heterogeneity of tight junction complexity between different segments of the vascular tree has been reported in other tissues. Similarly, contamination with endothelial cells derived from leaky circumventricular organs could also cause low TER *in vitro*.

In vitro models of the blood–brain barrier

Most brain endothelial cell culture protocols do not include any data on TER, but a few groups have achieved resistance levels which approach those measured *in vivo*. In this section four protocols developed by four different groups (Rutten *et al.*, 1987; Dehouck *et al.*, 1990; Rubin *et al.*, 1991; Raub *et al.*, 1992) will be compared with each other. They have in common that they have achieved brain endothelial cell monolayers *in vitro* which exhibit TER levels of between 600 and 700 Ω cm^2. Figure 4.5 shows flow diagrams of the isolation procedures and describes under what conditions TER measurements are taken.

In all protocols bovine cerebral cortices provide the starting material; however, the protocols differ from each other with respect to whether white and/or grey matter is used and possibly also with respect to the age of the animals used. After sampling the tissue, it is minced. Beyond this point the protocols differ considerably from each other.

In three protocols (Rutten *et al.*, 1987; Dehouck *et al.*, 1990; Rubin *et al.*, 1991) mincing of the tissue is followed by tissue homogenisation, whereas in the remaining protocol it undergoes the first of two enzymatic digest steps. One protocol does not include any enzymatic digest at all (Dehouck *et al.*, 1990), whereas two contain one digest (Rutten *et al.*, 1987; Rubin *et al.*, 1991) and the remaining one a succession of two digests (Raub *et al.*, 1992). The type of enzymes used, their concentration and the incubation time are different in all cases.

The tissue is further mechanically broken up by either a two-step sucrose gradient centrifugation (Rutten *et al.*, 1987), by two filtration steps through nylon meshes of different pore sizes with an intercalated further homogenisation step (Dehouck *et al.*, 1990) or just one filtration step (Rubin *et al.*,

1991), or by dextran density centrifugation in combination with a Percoll gradient centrifugation step (Raub *et al.*, 1992).

Another major difference between all protocols is the presence or absence of glial cell derived influence. Dehouck and coworkers (1990) grow astrocytes on the bottom side of filters on which endothelial cells are seeded, but during the initial growth phase and during the isolation of individual clones no astrocyte influence is present. Rubin *et al.* (1991), in contrast, stress the importance of the continuous presence of astrocyte conditioned medium (ACM). Both groups use astrocytes derived from rat tissue. Raub and coworkers (1992) also have glial influence present throughout, but they use the rat-derived C6 glioma cell line and grow them in the bottom chamber of their two-chamber arrangement. Unlike these three protocols, the one by Rutten *et al.* (1987) does not include medium conditioned by glial cells or glial cell co-culture. Growth medium, growth supplements and serum and growth substrates are also different in all four protocols.

Whereas Rutten *et al.* (1987) and Dehouck *et al.* (1990) isolate clones and therefore use cells after several passages, Raub *et al.* (1992) use primary cells derived from fragments which have been plated directly onto permeable filters. Rubin and coworkers (1991) use cells after one passage; in addition, they raise the level of intracellular cAMP of their endothelial cell cultures by addition of a membrane-permeable cAMP analogue and a phosphodiesterase inhibitor.

Despite the existence of considerable procedural differences between the four protocols, all groups claim to have achieved TER levels of between 600 and 700 Ω cm^2, thus approaching those measured *in vivo*. No obvious common trend in the isolation procedure and cell culture is apparent that could easily explain the successful development of a relatively tight brain endothelial cell monolayer.

The potential role of astrocytes remains particularly puzzling. Although several recent studies have provided further support for the inductive influence of glial cells on brain endothelial cell differentiation, some of the protocols discussed here appear to contradict this notion. It has recently been shown by use of a different tissue culture model that the elevation of intracellular cAMP levels and the presence of astrocytes or astrocyte-conditioned medium increases the complexity of tight junctions and is effective in maintaining the P-face association of tight junctional particles as revealed by quantitative freeze-fracture electron microscopy (Wolburg *et al.*, 1994). Both high complexity of tight junctions and P-face association of particles are characteristic of the *in vivo* BBB and are rapidly lost *in vitro* (Wolburg *et al.*, 1994). This observation adds independent support to the notion that

AUTHORS:	Rutten et al. (1987)	Dehouck et al. (1990)	Rubin et al. (1991)	Raub et al. (1992)
ISOLATION PROCEDURE:	Grey matter of calf brain cortices ↓ Mince ↓ Homogenise ↓ Rinse ↓ Two-step sucrose gradient centrifugation ↓ Rinse pellet ↓ Digest, trypsin/collagenase, 5 min ↓ Sparsely plate onto collagen/fibronectin coated dishes ↓ Clone ↓ Use after several passages	Bovine cerebral cortices ↓ Mince ↓ Homogenise ↓ Filter through 180 μm nylon mesh ↓ Homogenise filtrate ↓ Collect on 60 μm nylon mesh ↓ Plate onto extracellular matrix secreted by bovine corneal endothelial cells ↓ Isolate clones, plate onto gelatin-coated dishes ↓ After three passages: store frozen	Grey matter of bovine cerebral cortices ↓ Mince ↓ Homogenise ↓ Filter through 155 μm nylon mesh ↓ Digest, trypsin/collagenase, 60 min ↓ Plate onto collagen/fibronectin coated flasks ↓ (Alternatively: clone)	(Isolation according to Audus & Borchard, 1986) ↓ Grey matter of bovine cerebral cortices ↓ Mince ↓ Digest dispase 3 h ↓ Centrifuge ↓ Dextran density centrifugation of pellet ↓ Digest pellet, collagenase/dispase 5 h ↓ Centrifuge ↓ Percoll gradient centrifugation of pellet ↓ Collect microvessels ↓ Store frozen

GROWTH MEDIUM:	DMEM, 5% FCS, 5% NuSerum	DMEM, 15% FCS, 1 ng/ml bFGF	50% ACM (10% FCS), 50% MEM (10% PDHS), 10 ng/ml bFGF, 123 µg/ml heparin	
EXPERIMENTAL CONDITIONS FOR TER MEASUREMENTS: CELLS ARE GROWN ON MICROPOROUS FILTERS:	• Grow on specially designed porous filter coated with type I collagen and fibronectin • Medium: DMEM, 5% FCS, 5% NuSerum • Monolayers confluent within 7–12 days	• Coat 0.4 µm pore size filters on both sides with rat-tail collagen • Plate astrocytes on bottom side of filters • After 8 days, seed endothelial cells which have been passaged once after thawing onto top side • Medium: DMEM, 15% FCS, 1 ng/ml bFGF • Confluent after 8 days	• Trypsinise confluent monolayers of primary endothelial cells and densely plate onto 0.4 µm pore sized polycarbonate filters which are rat-tail collagen and fibronectin coated • Medium: 50% ACM (10% FCS), 50% N2 defined medium (Bottenstein & Sato, 1979) • After 2–3 days: treat with 250 µm CPT-cAMP, 17.5 µM phosphodiesterase inhibitor RO20-1724 to elevate intracellular cAMP level • Use after 2–3 days	• Plate thawed fragments onto 0.4 µm pore sized polycarbonate filters which are rat-tail collagen and fibronectin coated • Medium: 45% MEM, 45% F12 nutrient mixture (Ham), 10% PDHS, 100 µg/ml heparin • Transfer to co-culture arrangement (C6 rat glioma cells grown in lower chamber) at day 8 • Use after 1–2 days
TER:	Spontaneous TER: 160–780 $\Omega\,cm^2$ (depending on clone) Stable TER for several passages in culture	Spontaneous TER: 420 $\Omega\,cm^2$ TER after 1 week in astrocyte co-culture: 660 $\Omega\,cm^2$	Spontaneous TER (without ACM): 60 $\Omega\,cm^2$ TER with ACM: 120 $\Omega\,cm^2$ TER with ACM after cAMP-treatment: 600 $\Omega\,cm^2$	Spontaneous TER: $\leq 160\ \Omega\,cm^2$ TER with C6 glioma cell co-culture: 4.4-fold increase

Fig. 4.5. Comparison of four isolation procedures for microvascular brain endothelial cells yielding high transendothelial electrical resistance (TER). DMEM, Dulbecco Modified Eagle Medium; FCS, fetal calf serum; bFGF, basic fibroblast growth factor; ACM, astrocyte conditioned medium; PDHS, plasma-derived horse serum.

astrocyte influence and high levels of the second messenger cAMP may be important in maintaining differentiation of brain endothelial cells in culture. However, it appears to be possible to achieve high levels of TER in the absence of glial influence (Rutten *et al.*, 1987).

Comparison of the four protocols suggests that fairly different approaches can lead to similar results. However, it is difficult to explain why many other groups have apparently failed to grow well-differentiated, tight brain endothelial cell monolayers in culture by use of very similar procedures.

Reproducibility

An explanation of the problems associated with reproducibility of *in vitro* models of the BBB may be that vital aspects of the isolation procedures and growth conditions are not transmitted in sufficient detail. It could, for example, be of great importance that where tissue is homogenised, exact details about brand, size and clearance of the homogeniser as well as the volume of tissue and medium added and the speed and number of strokes are given. Similarly, general procedures such as the need for serum batch testing, batch testing of growth factors and even testing of enzymes for passaging may be particularly important for this type of cell. Choice and preparation of growth substrate is also likely to have a strong influence on the phenotype of the cells. Indeed, experience in our own laboratory, where we have further developed the method of Rubin *et al.* (1991), has shown that minute details in the way these cells are passaged, critically determine the level of TER achieved (L. Morgan, A. Charalambou and L.L. Rubin, unpublished data). In addition, initial plating density of capillary fragments and duration of the growth phase before seeding of the endothelial cells onto permeable filters were found to be critical.

Conclusion

Whereas numerous protocols for the growth of brain endothelial cells in culture are now available for a range of different species, the establishment of an *in vitro* model of the BBB has proved very difficult. A few groups have succeeded in growing high-resistance brain endothelial cell monolayers, but reproduciblity has remained more difficult than anticipated (Rubin, 1991). Research into the role of astrocytes or other elements of the brain environment on the differentiation of BBB endothelial cells and its implication for the pursuit of an *in vitro* model has not significantly progressed in the last few years. However, first signs are emerging from our laboratory that problems relating to reproducibility of *in vitro* models of the BBB can be overcome.

Acknowledgements

I am grateful to Jo Brashaw and Louise Morgan for critically reading the manuscript.

References

Audus, K.L. & Borchard, R.T. (1986). Characterization of an *in vitro* blood–brain barrier model system for studying drug transport and metabolism. *Pharm. Res.*, **3**, 81–7.

Bottenstein, J.E., & Sato, G.H. (1979). Growth of a rat neuroblastoma cell line in serum-free supplemented medium. *Proc. Natl. Acad. Sci. USA*, **76**, 514–17.

Chesné, C., Dehouck, M.-P., Jolliet-Riant, P., Brée, F., Tillement, J.-P., Dehouck, B., Fruchart, J.C. & Cecchelli, R. (1993). Drug transfer across the blood–brain barrier: comparison of *in vitro* and *in vivo* models. In *Frontiers in Cerebral Vascular Biology: Transport and its Regulation*, ed. L.R. Drewes & A.L. Betz, pp. 113–16. New York: Plenum Press.

Dehouck, M.-P., Méresse, S., Delorme, P., Fruchart, J.-C. & Cecchelli, R. (1990). An easier, reproducible, and mass-production method to study the blood–brain barrier *in vitro*. *J. Neurochem.*, **54**, 1798–1801.

Dehouck, M.-P., Jolliet-Riant, P., Brée, F., Fruchart, J.-P., Cecchelli, R. & Tillement, J.-P. (1992). Drug transfer across the blood–brain barrier: correlation between *in vitro* and *in vivo* models. *J. Neurochem.*, **58**, 1790–7.

Dermietzel, R. & Krause, D. (1991). Molecular anatomy of the blood–brain barrier as defined by immunocytochemistry. *Int. Rev. Cytol.*, **127**, 57–109.

Greenwood, J. (1991). Astrocytes, cerebral endothelium, and cell culture: the pursuit of an *in vitro* blood–brain barrier. *Ann. N.Y. Acad. Sci.*, **633**, 426–31.

Janzer, R.C. & Raff, M.C. (1987). Astrocytes induce blood–brain barrier properties in endothelial cells. *Nature*, **325**, 253–7.

Joó, F. (1992). The cerebral microvessels in culture: an update. *J. Neurochem.*, **58**, 1–17.

Laterra, J.J. & Goldstein, G.W. (1992). The blood–brain barrier *in vitro* and in culture. In *Physiology and Pharmacology of the Blood–Brain Barrier*, ed. M.W.B. Bradbury, pp 419–37. Berlin: Springer.

Laterra, J.J. & Goldstein, G. W. (1993). Brain microvessels and microvascular cells *in vitro*. In *The Blood-Brain Barrier: Cellular and Molecular Biology*, ed. W.M. Partdridge, pp. 1–24. New York: Raven Press.

Lillien, L.E., Sendter, M., Rohrer, H., Hughes, S. M. & Raff, M.C. (1988). Type-2 astrocyte development in rat brain cultures is initiated by a CNTF-like protein produced by type-1 astrocytes. *Neuron*, **1**, 485–94.

Partdridge, W.M., Triguero, D., Yang, J. & Cancilla, P.A. (1990). Comparison of *in vitro* and *in vivo* models of drug transcytosis through the blood–brain barrier. *J. Pharmacol. Exp. Ther.*, **255**, 893–9.

Raub, T.J., Kuentzel, S.L. & Sawada, G.A. (1992). Permeability of bovine brain microvessel endothelial cells *in vitro*: barrier tightening by a factor released from astroglioma cells. *Exp. Cell Res.*, **199**, 330–40.

Rubin, L.L. (1991). The blood–brain barrier in and out of cell culture. *Curr. Opin. Neurobiol.*, **1**, 360–3.

Rubin, L.L., Hall, D.E., Porter, S., Barbu, K., Cannon, C., Horner, H.C., Janatpour, M., Liaw, C.W., Manning, K., Morales, J., Tanner, L.I., Tomaselli, K.J. & Bard, F. (1991). A cell culture model of the blood–brain barrier. *J. Cell Biol.*, **115**, 1725–35.

Rutten, M.J., Hoover, R.L. & Karnovsky, M.J. (1987). Electrical resistance and macromolecular permeability of brain endothelial monolayer cultures. *Brain Res.*, **425**, 301–10.

Stewart, P.A. & Wiley, M.J. (1981). Developing nervous tissue induces formation of blood–brain barrier characteristics in invading endothelial cells: a study using quail–chick transplantation chimeras. *Dev. Biol.*, **84**, 183–92.

Takakura, Y., Audus, K.L. & Borchardt, R.T. (1991). Blood-brain barrier: transport studies in isolated brain capillaries and in cultured brain endothelial cells. *Adv. Pharmacol.*, **22**, 137–65.

Wolburg, H., Neuhaus, J., Kniesel, U., Krauss, B., Schmid, E.-M., Öcalan, M., Farrell, C. & Risau, W. (1994). Modulation of tight junction structure in blood–brain barrier endothelial cells: effects of tissue culture, second messengers and cocultured astrocytes. *J. Cell Sci.*, **107**, 1347–57.

5

Isolation, culture and properties of microvessel endothelium from human breast adipose tissue

Peter W. Hewett and J. Clifford Murray

Introduction

That endothelial cells derived from different vascular organs, and from within different vascular beds within those organs, display morphological, biochemical and antigenic heterogeneity is now beyond dispute (Kumar *et al.*, 1987; Zetter, 1988; Kuzu *et al.*, 1992; Hewett & Murray, 1993*a*). This fact has highlighted the need for methods to isolate and maintain in culture endothelial cells from a variety of species and tissues. Unquestionably the richest sources of microvascular endothelial cells have been tissues such as adipose tissues and brain which, by virtue of the special physical characteristics of their parenchyma, allow the ready separation of microvessels from the bulk of the tissue. The difference in buoyant densities of adipocytes and the stromal component of adipose tissues was first exploited by Wagner & Matthews (1975) for the isolation of microvessel endothelium from the rat epididymal fat pad. Similarly, high microvessel density, ease of separation, and the resistance of the parenchyma to growth in routine culture, have made brain a relatively accessible source of microvessel endothelium (Kumar *et al.*, 1987; Hewett & Murray, 1993*a*). Nevertheless the isolation of endothelial cells from more complex tissues with a wide variety of cell types, many of which readily adapt to cell culture in competition with the endothelial cell of interest, has until recently remained a significant technical problem. The development of new techniques producing relatively large numbers of pure endothelial cells has in addition led to the realisation that endothelial cells are less fastidious in terms of growth requirements than previously thought, and microvessel cells from a wide variety of tissues can now be routinely cultured.

Probably the most significant advance in the isolation of endothelial cells since the description of the isolation of human umbilical vein endothelial cells (HUVEC) (Jaffe *et al.*, 1973) is the use of superparamagnetic beads coupled to endothelial-specific ligands to select the endothelial cells from complex mixtures of cells. In the original procedure Jackson and colleagues (1990) used *Ulex europaeus* lectin (UEA-1), which binds with high affinity and specificity to α-fucosyl residues of glycoproteins associated with the endothelial cell surface (Holthöfer *et al.*, 1982). These determinants are also found in the H blood group antigen, although this does not usually present a significant problem in terms of culture contamination. We (Hewett & Murray, 1993*b*) recently refined this technique by coupling commercially available monoclonal antibodies directed against PECAM-1/CD31 (Newman *et al.*, 1990), a pan-endothelial marker (Kuzu *et al.*, 1992) to superparamagnetic beads, and now use this reagent routinely to isolate microvessel endothelial cells from a variety of sources (Hewett and Murray, 1993*b*).

Problems exist in terms of endothelial cell identity, even within the limited context of adipose tissue preparations. Several authors have reported the isolation of microvessel endothelium from human omental adipose tissue. However, there remains considerable controversy concerning the origin of these cells, and the weight of evidence would suggest that in many cases these cells are of mesothelial rather than endothelial origin (reviewed by Hewett & Murray, 1993*a*). Neither on morphological grounds nor on the basis of several conventional endothelial cell markers alone is it possible to distinguish these two cell types. We have suggested that the absence of PECAM-1 and E-selectin (endothelial cell leucocyte adhesion molecule-1, ELAM-1) (Bevilacqua *et al.*, 1987) expression by human mesothelial cells represents a suitable means of distinguishing these cells from endothelial cells (Hewett & Murray, 1994). Therefore the use of anti-PECAM-1 coated magnetic beads in principle obviates the need to use tissues devoid of mesothelium. Of course whichever technique is ultimately used, the problem of mesothelial cell contamination is negated by the use of adipose tissue from sites such as the breast, where there is no associated mesothelium.

In this chapter we describe the use of superparamagnetic beads coupled to anti-PECAM-1 to isolate microvessel endothelial cells from human breast adipose tissue obtained at reduction mammoplasty (Hewett *et al.*, 1993), and methods for the routine culture and characterisation of these cells. We also describe some of their functional and biochemical properties, and compare them with cells obtained from umbilical vein (Jaffe *et al.*, 1973) and other tissues.

Isolation of human mammary microvessel endothelial cells (HuMMEC)

Below we describe the use of 'PECA beads' to purify microvessel endothelial cells from breast adipose tissue digests. In addition this procedure has been used to isolate microvessel endothelial cells from omental and subcutaneous adipose tissues (Hewett & Murray, 1993*b*). Although we have found PECA beads to be more reliable for this purpose (Hewett & Murray, 1993*b*) purification can be achieved using Dynabeads coated with UEA-1 (Hewett *et al.*, 1993).

Materials

Preparation of magnetic beads coated with anti-PECAM-1 Anti-PECAM-1 coated beads (PECA beads) are prepared by incubation of equal volumes of goat anti-mouse IgG_1 secondary antibody (150 μg/ml) in 0.17 M sodium tetraborate buffer (pH 9.5) and sterile filtered with tosyl-activated superparamagnetic beads (Dynabeads M450; Dynal UK, Wirral, UK) for 24 h on a rotary stirrer at room temperature. Wash the beads four times for 10 min, and then overnight in sterile calcium- and magnesium-free Dulbeccos phosphate-buffered-saline (PBS/A; Difco) containing 0.1% bovine serum albumin (BSA) (PBS/A+0.1% BSA) on a rotary stirrer at 4 °C. (It is now possible to purchase Dynabeads pre-coated with secondary antibodies which are convenient and should work equally well.) We have used two monoclonal antibodies (MoAb) against PECAM-1 for the preparation of PECA beads: 9G11 (R & D Systems, Abingdon, Oxon, UK) and L133.1 (Becton-Dickinson, San Jose, CA) with similar results. Dilute the anti-PECAM-1 MoAb in PBS/A+0.1% BSA to give a final concentration of 10–20 μg per 10 mg of beads and incubate on a rotary stirrer for 16 h; free antibody is then removed by washing (as above). PECA beads will maintain their activity for more than 6 months when stored sterile at 4 °C. However, it is necessary to wash the beads with PBS/A+0.1% BSA prior to use to remove any free antibody that may be present.

Collagenase solution Dissolve type II collagenase (Sigma) at 2000 U/ml in Hank's Balanced Salt Solution (HBSS) containing 0.5% BSA, sterile filter, aliquot and store frozen (−20 °C).

Trypsin/EDTA solution Dilute stock trypsin solution (Sigma) in PBS/A to give 0.25% solution, add 1 mM (0.372 g/l) EDTA, ensure that it has dissolved, sterile filter, aliquot and store frozen (−20 °C).

Gelatin solution Dilute 2% stock gelatin solution (Sigma) in PBS/A to give a 0.2% solution and store at 4 °C. To coat tissue culture dishes add 0.2% gelatin solution and incubate for 1 h at 37 °C or overnight at 4 °C. Remove the gelatin solution immediately prior to plating the cells.

10% BSA solution Dissolve 10 g of BSA in 100 ml of PBS/A, filter sterilise and store at 4 °C.

100 μm nylon filters Cover the top of a polypropylene funnel (10 cm or greater diameter) with 100 μm nylon mesh filter (Lockertex, Warrington, Cheshire, UK) and sterilise by autoclaving.

Growth medium M199 with Earle's salts supplemented with 14 ml/l of 1 M *N*-[2-hydroxyethyl]piperazine-*N′*-[2-hydroxy-propane]sulphonic acid (HEPES) solution, 20 ml/l of 7.5% sodium hydrogen carbonate solution, 20 ml/l 200 mM L-glutamine solution, 20 ml of 100 U/ml penicillin, 100 mg/ml streptomycin solution, 1500 U/l of heparin (Leo Laboratories, Princes Risborough, Bucks, UK), 30% iron-supplemented calf serum (Hyclone, Logan, UT) and 40 μg/ml endothelial cell growth supplement (ECGS, Advanced Protein Products, Brierley, UK). Store at 4 °C for no more than 1 month.

Isolation of HuMMEC

Collect human breast adipose tissue obtained at reduction mammoplasty in a suitable large sterile container. The fat can be processed immediately or stored for up to 48 h at 4 °C. Place the tissue on a large sterile dish (Nunc bioassay dish, Nunc, IL) and wash with 2% antibiotic/antimycotic solution (Sigma) in PBS/A. Avoiding areas of dense (white) connective tissue, scrape the fat free from connective tissue fibres and large (visible) blood vessels with scalpels. Chop the fat up finely and place 10–20 g into a sterile 50 ml centrifuge tube. Add 10 ml of PBS/A and 5–10 ml of type II collagenase solution, shake the tube vigorously to break up the fat further and incubate with end-over-end mixing on a rotary stirrer at 37 °C for approximately 1 h.

Centrifuge the digests to separate the floating adipocytes and released oil from the aqueous lower layer containing the stromal components. Discard the fatty (top) layer and retain the cell pellet with some of the lower (aqueous) layer, add PBS/A and recentrifuge. Resuspend the cell pellet in 10% BSA solution and centrifuge (200 *g*, 10 min). Discard the supernatant, repeat the centrifugation in BSA solution and finally wash the pellet with 50 ml of PBS/A. Resuspend the pellet obtained in 0.25% trypsin/1mM EDTA solution for approximately 10–15 min with occasional agitation at 37 °C. Neutralise the trypsin solution by the addition of HBSS containing 5% calf serum (HBSS+5% CS) and mix thoroughly. Filter the suspension through 100 μm sterile nylon mesh to remove large fragments of connective tissue. Centrifuge the filtrate, and resuspend the resulting pellet in approximately 1–2 ml of ice-cold HBSS+5% CS.

Add approximately 50 μl of PECA beads and incubate for approximately 20 min at 4 °C with occasional agitation to facilitate attachment of the Dynabeads. It is important to keep the cell–Dynabead suspension cold during the purification steps to minimise non-specific phagocytosis of Dynabeads. Add HBSS+5% FCS to give a final volume of 10–12 ml and select the microvessel fragments using the magnetic particle concentrator (MPC-1; Dynal, Wirral, UK) for 3 min. Resuspend, wash and select the microvessel fragments a further 3–5 times in 12 ml of HBSS+5% FCS.

Finally, suspend selected cells in growth medium and plate onto 0.2% gelatin coated 25 cm^2 tissue culture flasks. Maintain cells at 37 °C in an atmosphere of 5% CO_2, changing the medium every 3–4 days. When confluent, HuMMEC are routinely passaged with trypsin/EDTA, onto gelatin coated dishes at a split ratio of 1:4.

Cryopreservation of cells

HuMMEC can be stored frozen in growth medium containing 10% (v/v) dimethyl sulphoxide in suitable cryovials. Cells ($\sim 2 \times 10^6$/ml) are cooled to −80 °C at 1 °C/min prior to storage in liquid nitrogen. We have not observed a loss of viability with HuMMEC stored in this manner for over 3 years.

HuMMEC in culture

Following the PECA bead selection procedure, small microvessel fragments and single cells coated with Dynabeads can be seen under light microscopy. After 24 h the cells flatten out to form distinct colonies (Fig. 5.1). HuMMEC isolated using the PECA bead technique grow to confluence within 10–14 days depending on the initial seeding density, forming

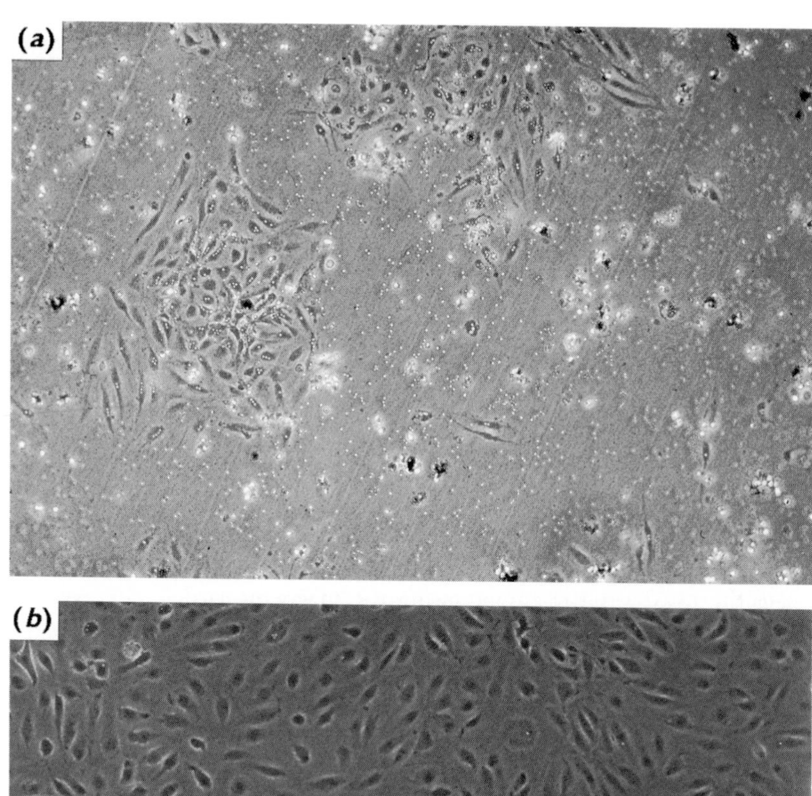

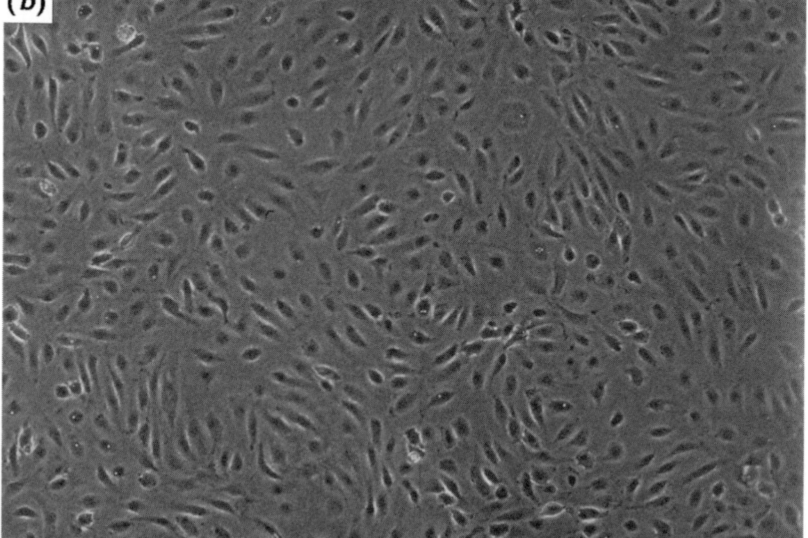

Fig. 5.1. Photomicrographs of HuMMEC: (*a*) cells growing out from microvessel fragments forming discrete colonies at 48 h after isolation and (*b*) confluent monolayers of HuMMEC (at passage 2) demonstrating typical cobblestone morphology.

cobblestone contact-inhibited monolayers (Fig. 5.1). The cobblestone morphology of HuMMEC is very typical of endothelial cells derived from many tissues and cells are easily distinguished from the major fibroblastoid contaminating cell populations (MCP) (Fig. 5.1).

We have successfully cultured these cells to passage 8 without observable change in morphology, but life-span is very much dependent on the individual preparation. We routinely use the cells between passages 3 and 6. Reselection with PECA beads and minor manual 'weeding' performed under under sterile conditions in a flow hood using a light microscope can be employed at passage 1 or 2 to maintain pure cultures. Reselection of cells is achieved by incubation of the cells with PECA beads following trypsinisation from flasks as described above. The formation of capillary-like tubes has been used as a means of discriminating between endothelial and mesothelial cells (reviewed by Hewett & Murray, 1993*a*). When cultured on Matrigel coated chamber slides HuMMEC will form tube-like networks within 4 h of plating (Fig. 5.2).

Effect of adherent PECA beads on endothelial cells in culture

We do not remove Dynabeads following cell selection, although this should be possible using Detachabead (Dynal). The Dynabeads are internalised within 24 h of selection and are diluted to negligible numbers per cell by the first passage, through cell proliferation (Jackson *et al.*, 1990). Polyclonal anti-PECAM-1 (EndoCAM) antibodies have previously been reported to inhibit the formation of confluent monolayer cells of bovine aortic endothelial cells. Consistent with previous observations with UEA-1 coated Dynabeads (Jackson *et al.*, 1990; Hewett *et al.*, 1993), PECA bead selection does not have any adverse effect on the adherence, proliferation (Fig. 5.3) or cobblestone morphology of endothelial cells (Hewett & Murray, 1993*b*).

Characterisation and functional properties of HuMMEC

There are many different morphological, ultrastructural, biochemical and immunological criteria on which identification may be based (Ruiter *et al.*, 1989). Over recent years, new and more specific endothelial markers have emerged, such as PECAM-1 and E-selectin (endothelial leucocyte adhesion molecule-1, ELAM-1). As many of these markers are not exclusive to endothelial cells several may be required to confirm endothelial identity. It is often advisable to demonstrate the absence of markers, such as smooth muscle α-actin (stress-fibres) and the intermediate filament protein, desmin,

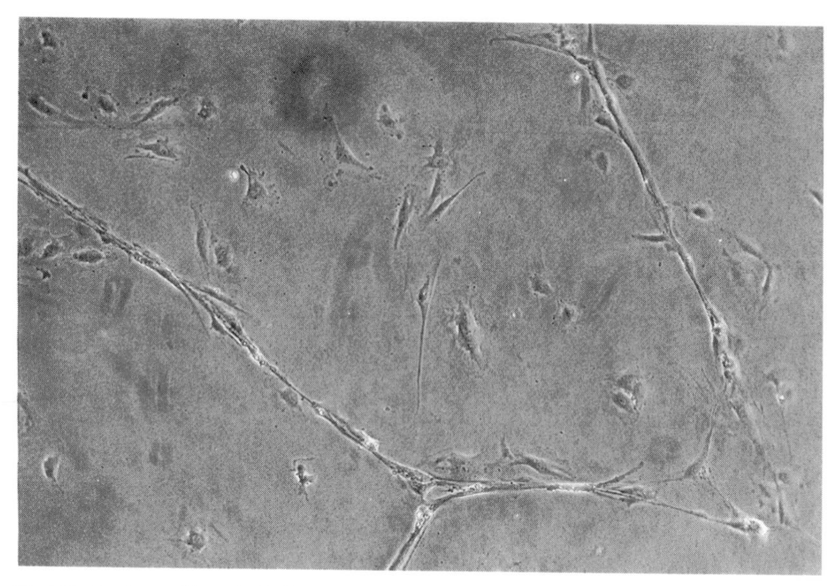

Fig. 5.2. Capillary-like tubule formation demonstrated by HuMMEC after 3 h of plating onto Matrigel.

which are specific for common contaminants such as pericytes and smooth muscle cells. It should be noted that endothelial cells demonstrate heterogeneity in expression of cell markers and lack of a given endothelial marker does not necessarily preclude the endothelial origin of isolates (Kumar *et al.*, 1987).

Immunocytofluorescence

As many antibodies against endothelial markers are commercially available (Table 5.1), immunocytofluorescence represents a simple technique with which to characterise endothelial cells. Outlined below is a standard protocol which can be used for this purpose.

Cell preparation Lab-Tek multiwell chamber slides are extremely useful for this purpose as several tests can be performed on the same slide using relatively few cells. Cells are cultured on glass chamber slides (Nunc, Naperville, IL) pre-treated for 1 h with 5 μg/cm^2 bovine fibronectin (Sigma) or 0.2% gelatin in normal growth medium. When sufficient cells are present the medium is discarded and cells washed twice with PBS/A, fixed in cold acetone (−20 °C, 10 min), air dried and stored frozen.

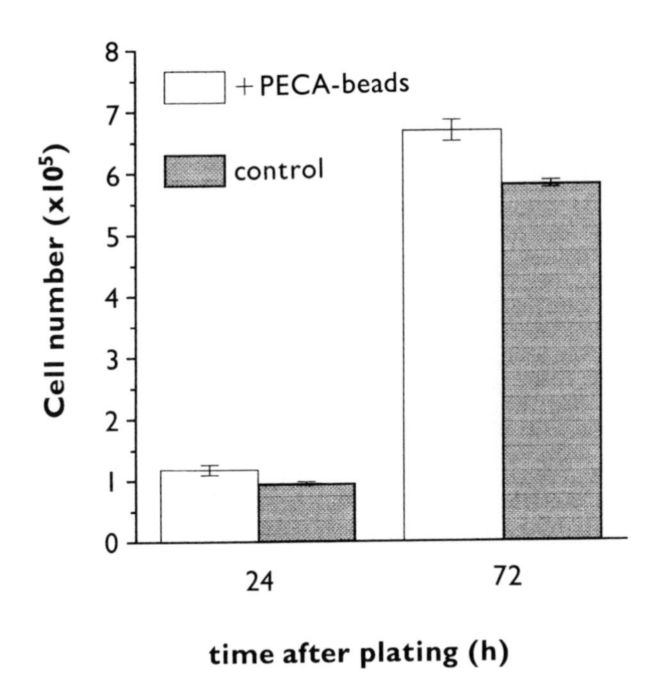

Fig. 5. 3. The effect of PECA beads on the proliferation of endothelial cells. HUVEC were harvested with trypsin/EDTA, washed with HBSS+5% FCS and incubated with 30 μl of PECA beads and purified as described in Materials. Selected and control (not incubated with Dynabeads) cells were plated onto gelatin coated, 6 well dishes and cell number determined at 24 and 72 h.

Immunocytofluorescence Allow slides to warm up from frozen to room temperature and wash with PBS/A (2×5 min). To prevent non-specific binding of the secondary antibody, block slides for 20 min with 10% normal serum from the same species in which the secondary antibody was raised. Incubate slides for 60 min with primary antibody applied after the recommended dilution in PBS/A (Table 5.1). After washing with PBS/A (3×5 min) incubate slides for 30 min with the appropriate fluorescein isothiocyanate (FITC)-labelled secondary antibody at a 1:50 dilution in PBS/A. Wash slides three times in PBS/A, mount coverslip with 50% (v/v) glycerol in PBS/A and observe under fluorescence microscopy. Stained slides can be stored for several months in the dark at 4 °C.

Controls To preclude false positives produced by non-specific binding of secondary antibodies, always include a negative control, consisting of cells treated as described above but with PBS/A substituted for the primary

Table 5.1. *Monoclonal antibodies (MoAb) used for immunocytofluorescent characterisation of isolated cell cultures*

MoAb (source)	Class/subclass	Dilution/conc.	Species	Target/antigen
F8/86 (Dako, High Wycombe, Bucks, UK)	IgG mouse	1:100	Human	von-Willebrand Factor (vWF)
EN4 (Sera-lab, Crawley Down, Sussex, UK)	IgM mouse	1:25	Human/cat	Unknown epitope in cytoplasm and at cell surface of endothelial cells
PAL-E (Sera-lab)	IgG_{2a} mouse	1:2	Human/bovine/goat/rabbit	Unknown epitope in cytoplasm of endothelial cells
QBend 10 (Serotec, Kidlington, Oxford, UK)	IgG_1 mouse	1:200 (10 μg/ml)	Human	CD34 (haematopoietic progenitor antigen)
3.1.1 (Affinity Bioreagents, NJ, USA)	IgM mouse	1:1000	Human/bovine/mouse/canine	Angiotensin converting enzyme (ACE)
H4-7/33 (Biogenesis, Bournemouth, UK)	IgG_1 mouse	1:20	Human/bovine	Unknown epitope at cell surface of endothelial and mesothelial cells
13D5 (R & D Systems, Abingdon, UK)	IgG mouse	5 μg/ml	Human	E-selectin/endothelial leucocyte cell adhesion molecule (ELAM-1)
9G11 (R & D Systems)	IgG_1 mouse	5 μg/ml	Human	Platelet endothelial cell adhesion molecule-1 (PECAM-1/CD31)
L.133.1 (Becton-Dickinson, San Jose, CA, USA)	IgG mouse	5 μg/ml	Human	PECAM-1
D33 (Dako)	IgG_1 mouse	1:25	Human	Intermediate filament protein desmin found in muscle cells, pericytes and mesothelial cells
1A4 (Sigma, Poole, UK)	IgG_{2a} mouse	1:200	Human	Smooth muscle α-actin; smooth muscle cells and transitional (smooth muscle-like) pericytes

antibody. It is also advisable to use other cell types, such as fibroblasts or smooth muscle cells, and previously characterised endothelial cells as negative and positive controls respectively.

E-selectin. To examine the expression of E-selectin, cells must be incubated with either tumour necrosis factor α (TNFα) or interleukin-1 (IL-1) in growth medium for 4 h. We routinely use 10 ng/ml recombinant human TNFα. Stimulated cells should be compared with unstimulated controls incubated with anti-E-selectin MoAb as described above.

Detection of UEA-1 binding

Lectins have proven useful cell markers and UEA-1 lectin has been used as a marker of human endothelial cells (Holthöfer *et al.*, 1982). Its specific interaction with α-L-fucosyl containing glycoproteins at the luminal cell surface can be visualised by the use of biotinylated peroxidase and FITC-labelled lectin or anti-UEA-1 antibodies. The binding of UEA-1 can be determined very rapidly using FITC-conjugated or tetramethylrhodamine isothiocyanate (TRITC)-conjugated UEA-1 (Sigma). Incubate acetone-fixed cells with 10–100 μg/ml of FITC-conjugated UEA-1 for 1 h at room temperature, wash with PBS/A and mount as described above. Alternatively unlabelled UEA-1 may be used and binding detected with rabbit anti-UEA-1 polyclonal antibody at 1 in 100 dilution in PBS/A for 1 h followed by FITC-labelled goat anti-rabbit antibody at a 1 in 50 dilution for 30 min (see above). In each case a control should be included to ensure that UEA-1 is binding specifically to α-L-fucose residues. Control slides are incubated with UEA-1 that has been pre-treated with 1 M α-L-fucose (Sigma) for 1 h prior to addition to the cells.

Metabolism of acetylated low density lipoprotein

Detection of scavenger receptors for acetylated low density lipoprotein is rapid and simple to perform using low density lipoprotein coupled to the fluorescent carbocyanine dye (1,1′dioctadecyl-3,3,3′,3′-tetramethyl-indocarbocyanine perchlorate acetylated low density lipoprotein, DiI-Ac-LDL) (Voyta *et al.*, 1984). Incubate cells growing on microculture slides for 4 h in growth medium containing 10 mg/ml DiI-Ac-LDL (Biogenesis, Bournemouth, Dorset, UK) at 37 °C as described previously (Voyta *et al.*, 1984), fix in 3% formaldehyde solution for 30 min at room temperature, mount and observe by fluorescence microscopy.

'Capillary-like' tube formation

To examine the ability of the cells to produce tubular structures *in vitro*, culture cells on chamber slides coated with Matrigel (Becton-Dickinson), an extract of the Engelbreth–Holm–Swarm (EHS) murine sarcoma containing basement membrane components or collagen type-I gels and observe at regular intervals by light microscopy. Collagen type-I gels can be produced by mixing rat-tail collagen type I (Boehringer Mannheim) at 3 mg/ml in M199 basal salt solution containing 0.2 M HEPES buffer and 0.15% $NaHCO_3$, on ice. Dispense this solution into pre-chilled microculture slides and allow to set for 30 min at 37 °C. Adjust the pH of the gels (as indicated by phenol red colour change) with ammonia. Monitor the cells at regular intervals to assess the formation of 'capillary-like' tube structures.

Measurement of angiotensin-converting enzyme (ACE)

Endothelial cell angiotensin converting enzyme (ACE) expression is maintained in short-term culture and can be assessed both immunohistochemically and by the use of several assays employing synthetic substrates. The method described by Catravas & Watkins (1985) works well for this purpose. Cells are plated in triplicate at high density ($1–3\times10^5$/well) onto gelatin coated 24 well plates and allowed to attach overnight. The following day discard the medium and, as ACE is a serum component, wash the cell monolayer twice with Earle's Basal Salt Solution (EBSS) (without phenol red indicator) to remove any traces of serum. Incubate the cell monolayers for 30 min at 37 °C with 0.6 ml [^{3}H]benzoyl-phenylalanyl-alanyl-proline ([^{3}H]BPAP, 25 Ci/mmol) diluted in EBSS (0.1 mCi/ml). Withdraw aliquots (200 μl) in duplicate from each well at 30 min and transfer to scintillation vials containing 2.3 ml of 0.1N HCl to terminate the reaction. Set up four control vials with 0.6 ml of [^{3}H]BPAP and 5 ml of aqueous scintillant (Ecolite) to give the total activity of the substrate used. Invert vials 20 times and separate [^{3}H]BPAP from the unreacted substrate by 48 h of extraction in 2.5 ml of organic scintillant (0.4% Omnifluor in HPLC grade toluene). The level of activity in the organic phase can then be determined in a scintillation counter. Normalise ACE activity by subtracting the four control blank values (background) from the sample values. ACE activity can be expressed in units per 10^5 cells, where 1 unit equals the amount of enzyme required to metabolise 1% of [^{3}H]BPAP per minute at 37 °C.

Immunohistochemical characteristics of HuMMEC

Results of the immunocytofluorescence characterisation studies performed with a variety of antibodies (Table 5.1) on HuMMEC are summarised in Table 5.2 and Figs. 5.4–5.7. For a more detailed literature on these endothelial markers refer to Hewett & Murray (1993*a*) and Ruiter *et al.* (1989).

von Willebrand Factor (vWF) vWF is expressed at significant levels only in endothelial cells, megakaryocytes, and in human syncytiotrophoblasts. It forms large macromolecular complexes with factor VIII, and is stored in endothelial cell Weibel–Palade (WP) bodies. These rod-shaped organelles originate from the trans-Golgi apparatus and are present in large numbers of human endothelial cells isolated from large vessels (Jaffe *et al.*, 1973). However, WP bodies are frequently reported to be scarce or absent in capillary endothelial cells derived from various species *in vitro* (Wagner & Matthews, 1975). Several studies have indicated that vWF may not be an ideal marker for microvascular endothelium. Capillary endothelial cells of the human lung, kidney, adrenals, lymphatic and liver sinusoidal endothelium and tumours are reported to be either very weakly positive or negative for vWF in histological sections (Kumar *et al.*, 1987; Kuzu *et al.*, 1992). However, microvascular endothelial cells isolated from human kidney, dermis, synovium (Jackson *et al.*, 1990), decidua, heart, adipose tissue and brain have been reported to express vWF in perinuclear granules. Similarly, HuMMEC, and other microvessel endothelial cells derived from adipose tissue and lung, all demonstrate strong granular perinuclear immunofluorescence for vWF (Hewett & Murray, 1993*b*), (Fig. 5.4). The majority of evidence in the literature suggests that mesothelial cells do not express and store vWF in significant quantities (Hewett & Murray, 1993*a*). The adhesion molecule P-selectin (GMP-140/CD62), like vWF, is unique to WP bodies, platelet α-granules and megakaryocytes, and so should also represent a useful endothelial cell marker.

PECAM-1 The presence of PECAM-1 on HuMMEC is characterised by typical intense membrane fluorescence at points of cell–cell contact (Fig. 5.5). PECAM-1 is a member of the immunoglobulin gene superfamily and is constitutively expressed on the surface of endothelial cells ($>10^6$ molecules/cell), and to a lesser extent in platelets, granulocytes and a subpopulation of $CD8^+$ lymphocytes (Newman *et al.*, 1990). PECAM-1 is a homotypic cell–cell, cell–matrix adhesion molecule and may also amplify β1 integrin mediated adhesion. It is a superior marker to vWF as it is present

Table 5.2. *Results of immunocytofluorescent characterisation of HuMMEC*

MoAb	HUVEC	HuMMEC	HuAMEC	HuOMEC	HuOMC	MCP
vWF	++	++	++	++	−	−
ACE	+	++	++	++	++	−
PAL-E	+	+	+	ND	−	−
EN4	++	++	ND	ND	ND	ND
CD34	+(<20%)	+(<10%)	ND	ND	−	−
PECAM-1/CD31	++	++	++	++	−	−
E-selectin	+	++	++	ND	−	−
H4-7/33	+	+	+	+	+	−
Desmin	−	−	−	−	+[a]	−
Smooth muscle α-actin	−	−	−	−	−	−

Notes:

−, no MoAb staining; +, positive immunofluorescence; ++, strong immunofluorescence; ND, test not performed on these isolates. HUVEC, human umbilical vein endothelial cells; HuMMEC, human mammary microvessel endothelial cells; HuAMEC, abdominal subcutaneous adipose microvessel endothelial cells; HuOMEC, abdominal omental adipose microvessel endothelial cells; HuOMC, abdominal omental mesothelial cells; MCP, major contaminating population of 'fibroblastic' cells commonly found in HuMMEC preparations not selected by Dynabeads.

[a] Desmin staining present on cells plated at high density.

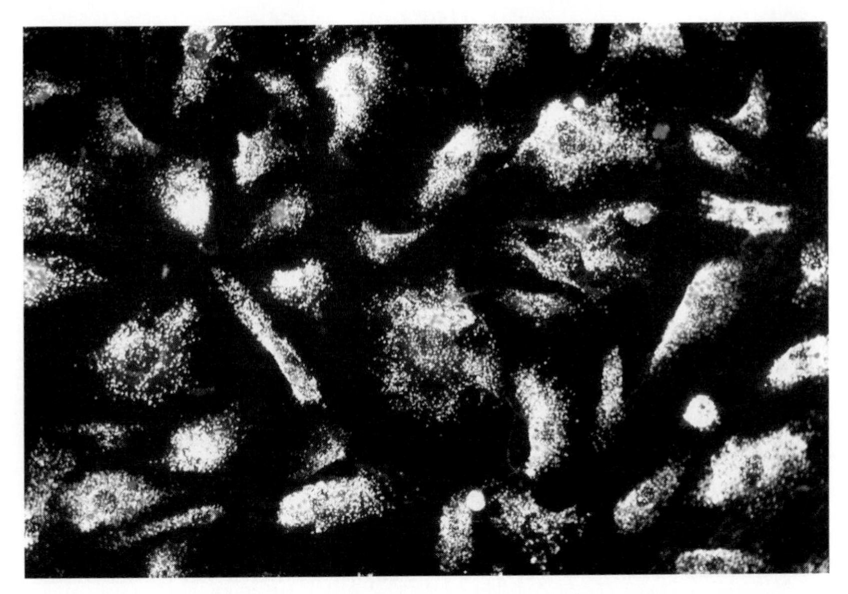

Fig. 5.4. Intense punctate perinuclear immunofluorescent staining of von Willebrand factor (vWF) in HuMMEC.

Fig. 5.5. Immunofluorescent staining of platelet endothelial cell adhesion molecule-1 (PECAM-1/CD31) in HuMMEC.

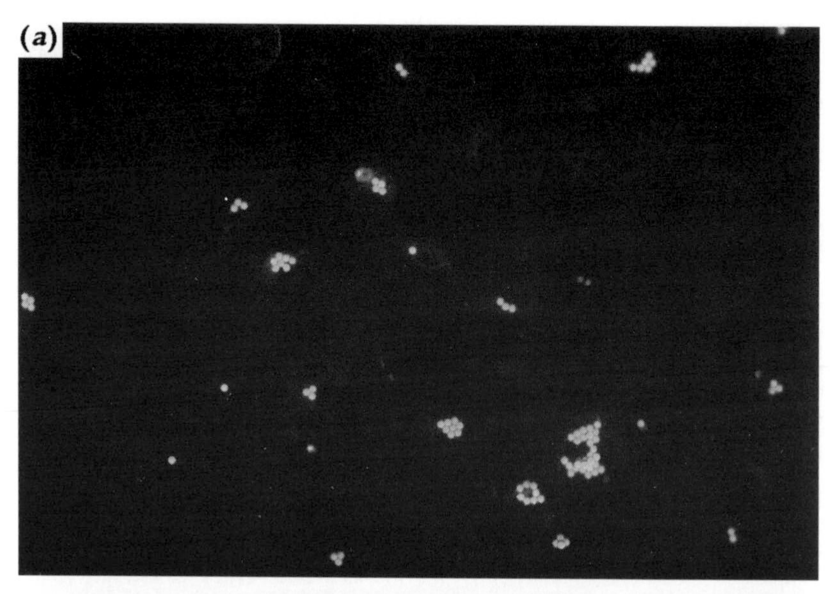

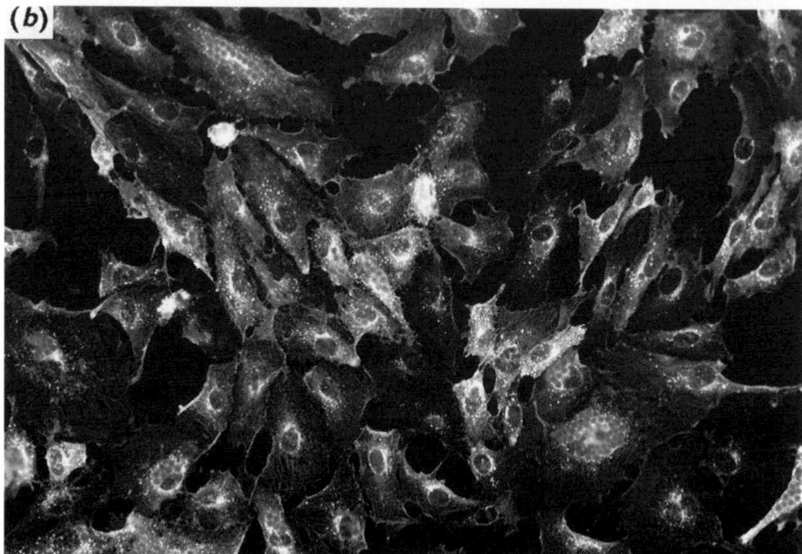

Fig. 5.6. HuMMEC incubated with the anti-E-selectin MoAb 13D5: (*a*) control, (*b*) after 4 h incubation with 1 ng/ml rh-TNFα.

on endothelia which by immunohistochemistry are vWF negative, notably lymphatic (high) endothelial cells, liver sinusoidal endothelium and renal glomerular capillaries (Kumar *et al.*, 1987; Kuzu *et al.*, 1992). In addition, MEC isolated from human adrenal cortex express PECAM-1 in the absence

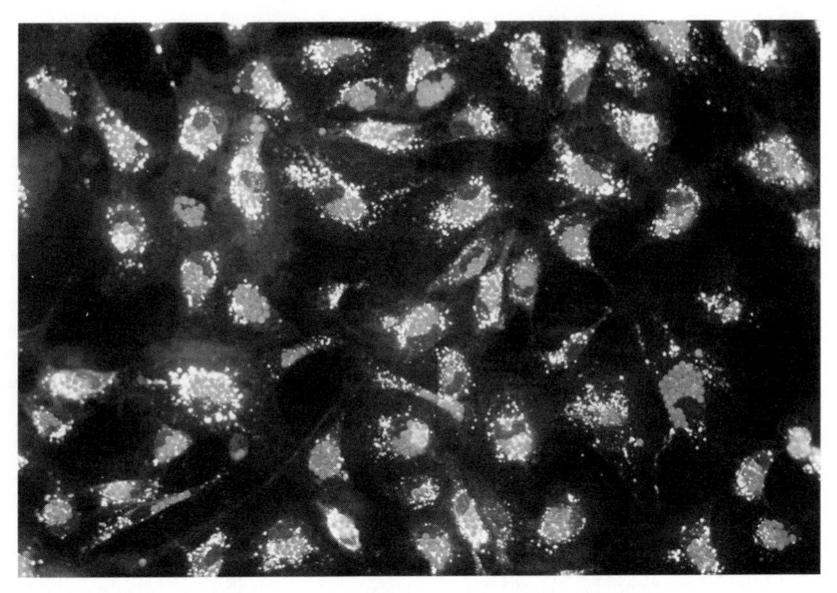

Fig. 5.7. Intense punctate fluorescence demonstrating the uptake of DiL-Ac-LDL in HuMMEC.

of vWF. PECAM-1 is not expressed by mesothelial cells in culture, and so is a good discriminatory test for contaminating mesothelial cells (Hewett & Murray, 1994).

EN4 EN4 was raised against HUVEC, and reacts with an unidentified antigen on all human endothelial cells (reviewed by Ruiter *et al.*, 1989). The EN4 antigen displays similar mass (130 kDa) to PECAM-1 and localises to the intercellular junctions of endothelial cells in culture, giving a similar pattern of cell staining and distribution to PECAM-1.

E-selectin Expression of E-selectin appears to be unique to endothelial cells (Bevilacqua *et al.*, 1987). Although not constitutively expressed by the majority of endothelial cells, following TNFα stimulation expression reaches a maximum at 4–8 h before returning to background levels within 24 h (Bevilacqua *et al.*, 1987). Absent on unstimulated controls, intense E-selectin expression is induced in HuMMEC by pretreatment for 4 h with as little as 100 pg/ml of TNFα (Fig. 5.6). E-selectin is a member of the selectin/LEC-CAM family of adhesion molecules thought to recognise sialyl-Lewis^x (SLe^x) and -Lewis^a (SLe^a), facilitating the binding of polymorphonuclear leucocytes and, to a lesser degree, lymphocytes to endothelial cells in inflammatory situations. E-selectin is not induced by TNFα treatment of human mesothelial cells in culture (Hewett & Murray, 1994).

PAL-E PAL-E reacts with an unidentified antigen associated with vesicles of human, bovine, cat, goat and rabbit endothelial cells and is a good marker of endothelium in many normal tissues and tumours (reviewed by Ruiter *et al.*, 1989). However, it will not stain all arterial endothelial cells and is not expressed in normal brain endothelium. PAL-E staining of HuMMEC is weak but similar to our observations of other endothelial cells (Table 5.2). PAL-E is not expressed by cultured mesothelial cells (Ruiter *et al.*, 1989).

QBend 10/anti-CD34 The haematopoietic progenitor antigen CD34 is expressed on 1–4% of human bone marrow cells and the luminal surface of both normal and pathological endothelium (Kuzu *et al.*, 1992; Schlingemann *et al.*, 1990). However, CD34 expression appears to be down-regulated in endothelial cells *in vitro.* Only a proportion (5–10%) of HuMMEC stain with QBend 10 and these cells appear to exhibit a migratory 'sprouting' phenotype situated above the plane of the cell monolayer in a similar manner to HUVEC (Schlingemann *et al.*, 1990).

H4-7/33 The MoAb H4-7/33 was raised against HUVEC and reacts with an unknown epitope common to all endothelial cells, with preference for the very small capillaries of the lung. It also binds to osteosarcoma and melanoma cell lines. Although, H4-7/33 reacts with HuMMEC and other endothelial cells it also produces similar staining of mesothelial cells (Hewett & Murray, 1994) (Table 5.2).

UEA-1 UEA-1 exhibits intense specific binding to the surface of HuMMEC. UEA-1 binds specifically to α-L-fucosyl-containing glycoproteins (Holthöfer *et al.*, 1982). Although UEA-1 has been reported not to bind to human omental mesothelium in histological sections, or mesothelial cells *in vitro*, we and others have found that UEA-1 will also bind specifically to other epithelial and mesothelial cells (Jaffe *et al.*, 1983; Hewett & Murray, 1994).

The major contaminating cell population (MCP) present in unselected cultures was also examined for expression of these endothelial markers, but did not bind any of the endothelial-specific MoAbs (Table 5.2). These cells do not contain desmin and smooth muscle α-actin, precluding the possibility that they are pericytes or smooth muscle cells. The use of medium containing high concentrations of D-valine (690 mg/ml) substituted for L-valine has previously been described as a means of inhibiting growth of fibroblasts in culture. However, use of this medium did not affect the proliferation of MCP in culture.

Functional characteristics of HuMMEC

DiI-Ac-LDL uptake Acetylated low density lipoprotein has been used not only for endothelial cell identification but for specific cell isolation (Voyta *et al.*, 1984). Endothelial cells possess scavenger receptors for this modified lipoprotein, whereas fibroblasts, pericytes and SMC do not. The DiI-Ac-LDL is taken up into secondary lysosomes where the low density lipoprotein is broken down and the dissociated lipophilic carbocyanine dye (DiI) is incorporated into the lysosomal membrane giving characteristic intense red punctate staining under fluorescence microscopy (Fig. 5.7). However, DiI-Ac-LDL is also taken up by macrophage and monocytes which have limited viability *in vitro*. The majority of reported evidence suggests that mesothelial cells are stained with DiI-Ac-LDL more intensely than endothelial cells (reviewed by Hewett & Murray, 1993*a*). However, other authors have reported scavenger receptors to be expressed at low levels or absent in mesothelial cells (reviewed by Hewett & Murray, 1993*a*, 1994).

ACE activity ACE is located at the luminal surface of endothelial cells and acts primarily as a dipeptidyl carboxypeptidase. Its main functions are the conversion of angiotensin I to its active form angiotensin II and the inactivation of kinins such as bradykinin. The ACE activity of HuMMEC isolated from three individuals is compared with that of microvessel endothelial cells derived from omental and subcutaneous adipose tissue, and human lung, HUVEC and omental mesothelial cells in Fig. 5.8. Microvessel endothelial cells expressed greater ACE activity than HUVEC. Although mesothelial cells have been reported to express high levels of ACE activity (reviewed in Hewett & Murray, 1993*a*) our results indicate lower activity in three mesothelial isolates. The major site of ACE activity is the microvasculature of the lung; this may be due either to increased expression of ACE *per se* in these vessels, or to the extensive surface area of the lung microvasculature. The expression of ACE has been reported to be much greater (3- to 5-fold) in human arterial endothelium than venular endothelium. The greater ACE activity expressed by microvessel endothelial cells compared with HUVEC may therefore be a consequence of relative position in the vascular network rather than an organ-specific difference. ACE does not show absolute specificity for endothelium and is expressed on the proximal convoluted tubules

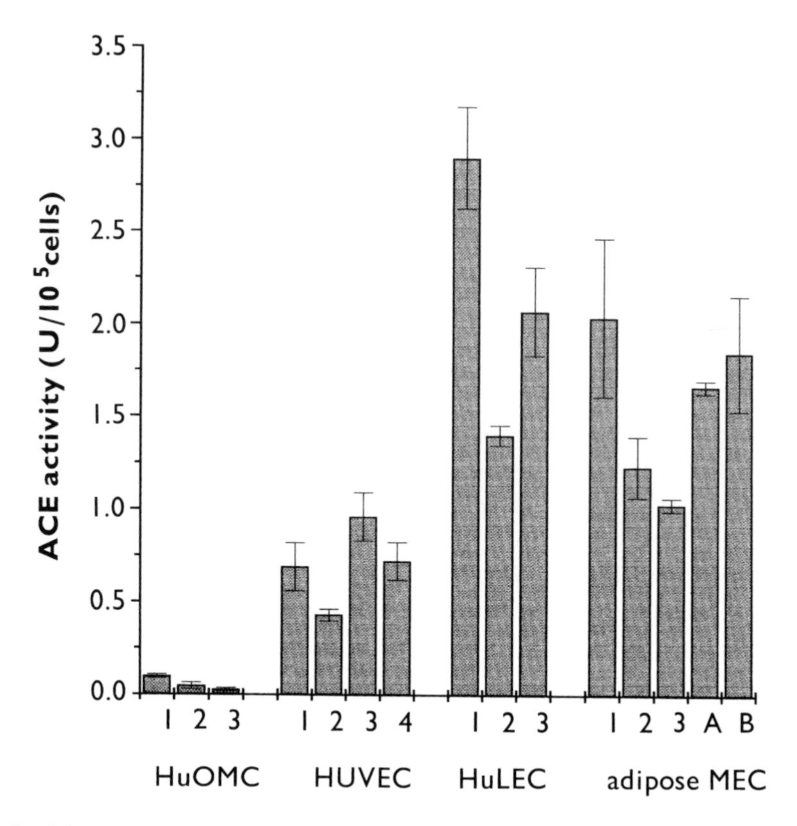

Fig. 5.8. Angiotensin converting enzyme (ACE) activity of mesothelial cell lines (HuOMC 1–3), in comparison with endothelial cells from human umbilical veins (HUVEC 1–4), microvessel endothelial cells from lung (HuLEC 1–3), and adipose MEC: mammary (HuMMEC 1–3), subcutaneous (HuAMEC A) and omental (HuOMEC B). Results represent the mean (±1 SD) of six determinations (1 unit=the amount of enzyme required to metabolise 1% [^{3}H]benzoyl-phenylalanyl-alanyl-proline ([^{3}H]BPAP) per minute at 37 °C).

of kidney, brush border membranes of absorptive epithelia (e.g. in small intestine), choroid plexus, macrophages, T lymphocytes and fibroblasts in culture.

Expression of the endothelial-cell-specific RTKs

Recently, there has been considerable interest in the receptor tyrosine kinases (RTKs) involved in endothelial cell differentiation and proliferation, particularly since the discovery and characterisation of vascular endothelial cell growth factor/vascular permeability factor (VEGF) (reviewed by Plate *et al.*, 1994). VEGF is a key angiogenic factor and is expressed in both physiological and pathological situations. It is unique in that it demonstrates pleiotropic activities specifically on endothelial cells, stimulating proliferation, inducing procoagulant activity and increasing vascular permeability. Two class (III) RTKs, KDR, its murine homologue Flk-1, and Flt-1 have recently been identified as high-affinity receptors for VEGF, and play a critical role in vasculogenesis and angiogenesis. We have found that HuMMEC in culture proliferate in response to VEGF and express *flt-1* and KDR at similar levels to other microvessel endothelial cells and HUVEC.

The endothelial RTKs *tie* and *tek/tie-2* have only recently been identified and together will form a new class (VIII) of RTKs which, as their ligand(s) remain obscure, have yet to be ascribed a function. Like the VEGF receptors these RTKs are expressed at high levels in endothelial cells during vasculogenesis and angiogenesis in the adult. By the reverse transcriptase polymerase chain reaction we have examined the expression of *tie* and *tek* in a range of different primary human cell cultures and have only detected them in endothelial cells. Similar levels of *tie* and *tek* were detected in HuMMEC and other microvessel and large vessel endothelial cells. As antibodies to these RTKs become available they should represent good markers for endothelial cell characterisation.

References

Bevilacqua, M.P., Pober, J.S., Mendrich, D.L., Cotran, R.S. & Gimbrone, M.A. (1987). Identification of an inducible endothelial-leukocyte adhesion molecule. *Proc. Natl. Acad. Sci. Usa*, **84**, 9238–42.

Catravas, J.D. & Watkins, C.A. (1985). Plasmalemmal metabolic activities in cultured calf pulmonary arterial endothelial cells. *Res Commun. Chem. Pathol. Pharmacol.*, **50**, 163–79.

Hewett, P.W. & Murray, J.C. (1993*a*). Human microvessel endothelial cells: isolation, culture and characterization. *In Vitro Cell Dev. Biol.*, **29A**, 823–30.

(1993*b*). Immunomagnetic purification of human microvessel endothelial cells using Dynabeads coated with monoclonal antibodies to PECAM-1. *Eur. J. Cell Biol.*, **62**, 451–4.

(1994) Human omental mesothelial cells: a simple method for isolation and discrimination from endothelial cells. *In Vitro Cell Dev. Biol.*, **30A**, 145–7.

Hewett, P.W., Murray, J.C., Price, E.A., Watts, M.E. & Woodcock, M. (1993). Isolation and characterization of microvessel endothelial cells from human mammary adipose tissue. *In Vitro Cell Dev. Biol.*, **29A**, 325–31.

Holthöfer, H., Virtanen, I., Kariniemi, A.-L., Hormia, M., Linder, E. & Miettinen,

A. (1982). *Ulex europaeus* 1 lectin as a marker for vascular endothelium in human tissues. *Lab. Invest.*, **47**, 60–6.

Jackson, C.J., Garbett, P.K., Nissen, B. & Schrieber, L. (1990). Binding of human endothelium to *Ulex europaeus* -1 coated Dynabeads: application to the isolation of microvascular endothelium. *J. Cell Sci.*, **96**, 257–62.

Jaffe, E.A., Nachman, R.L., Becker, C.G. & Minidi, C.R. (1973). Culture of human endothelial cells derived from umbilical veins: identification by morphological and immunological criteria. *J. Clin. Invest.*, **52**, 2745–56.

Kumar, S., West, D.C. & Ager, M. (1987). Heterogeneity in endothelial cells from large vessels and microvessels. *Differentiation*, **36**, 57–70.

Kuzu, I., Bicknell, R., Harris, A.M., Jones, M., Gatter, K.G. & Mason, D.Y. (1992). Heterogeneity of vascular endothelial cells with relevance to diagnosis of vascular tumors. *J. Clin. Pathol.*, **45**, 143–8.

Newman, P.J., Berndt, M.C., Gorski, J., White G.C., II, Lyman, S., Paddock, C. & Muller, W.A. (1990). PECAM-1 (CD31) cloning and relation to adhesion molecules of the immunoglobulin gene superfamily. *Science*, **247**, 1219–22.

Plate, K.H., Breier, G. & Risau, W. (1994). Molecular mechanisms of developmental and tumour angiogenesis. *Brain Pathol.*, **4**, 207–18.

Ruiter, D.J., Schlingemann, R.O., Rietveld, F.J.R & de Waal, R.M.W. (1989). Monoclonal antibody-defined human endothelial antigens as vascular markers. *J. Invest. Dermatol.*, **93**, 25S–32S.

Schlingemann, R.O., Rietveld, F.J.R., de Waal, R.M.W., Bradley, N.J., Skene, A.I., Davies, A.J.S., Greaves, M.F., Denekamp, J. & Ruiter, D.J. (1990). Leukocyte antigen CD34 is expressed by a subset of cultured endothelial cells and on endothelial abluminal microprocesses in the tumour stroma. *Lab. Invest.*, **62**, 690–6.

Voyta, J.C., Via, D.P., Butterfield, C.E. & Zetter, B.R. (1984). Identification and isolation of endothelial cells based on their increased uptake of acetylated-low density lipoprotein. *J. Cell Biol.*, **99**, 2034–40.

Wagner, R.C. & Matthews, M.A. (1975). The isolation and culture of capillary endothelium from epididymal fat. *Microvasc. Res.*, **10**, 286–97.

Zetter, B.R. (1988). Endothelial heterogeneity: influence of vessel size, organ localization and species specificity on the properties of cultured endothelial cells. In *Endothelial Cells*, vol. 2, ed. U.S. Ryan. Boca Raton, Florida: CRC Press.

6

Human skin microvascular endothelial cells

Zbigniew Ruszczak

Introduction

Local differentiation gives rise to remarkable endothelial heterogeneity in the vascular and lymphatic systems of different organs. Because of its strategic location, microvascular endothelium plays a central role in skin physiology, and its involvement in pathologies of the skin is becoming increasingly apparent. Thus, endothelial cells play a role in anti-thrombogenicity, blood vessel permeability, blood pressure control, metabolism of lipoproteins, tissue ageing, antigen presentation, and angiogenesis in wound healing. Dysfunctions are associated with a number of pathologies such as inflammation, immune disorders and hyperproliferation of vessels in, for example, psoriasis.

In vitro culture of endothelial cells from large vessels (aorta and umbilical vein) has provided a useful model for the study of endothelial cell metabolism; however, the validity of using endothelial cells isolated from the macrovasculature to study microvascular behaviour and function is questionable. Nearly 20 years ago, the group led by Karasek developed endothelial cell cultures derived from capillaries of rabbit skin (Davison & Karasek, 1978) and from the human newborn foreskin dermis. These established more representative *in vitro* models of the skin microvasculature (Davison *et al.*, 1980). More recently, Davison *et al.* have also published a method for the isolation and long-term serial cultivation of endothelial cells derived from microvessels of the adult human dermis (Davison *et al.*, 1983).

Cultured skin microvascular endothelial cells have a wide range of research applications including the study of cellular differentiation, cytokine expression, haemorrhagic disorders, hypersensitivity, inflammation, thrombogenesis and thrombosis, tumour metastasis, and wound healing. Because of a large reduction in toxicological tests involving laboratory animals, *in vitro* cell culture models are of increasing importance.

For these reasons, understanding of the physiology, biology and immunology of the capillary endothelium – the largest immunologically active organ in man – was and still is one of the most important areas in experimental medicine and biology.

The most difficult step in the preparation of primary human dermal microvascular cells is the isolation of the primary cell population for seeding. The development of cell culture media that support the growth of 'selected cell populations' (e.g. keratinocytes, fibroblasts and endothelial cells) has led to ease of culture, but careful separation of the cell types for seeding remains of paramount importance.

In 1989 we developed a simple gradient centrifugation technique for the separation of microvascular endothelial cells from other cell populations obtained from human skin isolates (Imcke *et al.*, 1989; Ruszczak *et al.*, 1990). The method was established not only for traditional sources of human microvascular endothelium such as newborns' and children's foreskin, but also for adult skin obtained from various localities, as well as vascular tumours such as angiomas and kaposi's sarcoma (Ruszczak *et al.*, 1990).

Background methods for the isolation and culture of dermal endothelium *in vitro*

The culture of endothelial cells started in 1922 with the experiments of Lewis, who described the first successful outgrowth of vascular cells from explants. As noted in the Introduction, the first successful isolation and culture procedure for human dermal microvascular endothelial cells (HDMEC) was published in 1980 by Davison and co-workers (Davison *et al.*, 1980). Newborn human foreskin, obtained following routine circumcision, was collected in 0.15 M NaCl containing 400 IU/ml penicillin and 200 μg/ml streptomycin and stored at 4 °C for up to 24 h before use. The epidermis was removed from the vascular reach region using a Castroviejo keratotome set to cut at 0.1 mm. After exposure to 0.3% trypsin solution containing 1% EDTA for 40 min at 37 °C the epidermis was removed, dermal fragments were washed several times in isotonic salt solution, and the foreskin vessels squeezed out with a scalpel blade into medium containing 10% pooled human serum. Vessels isolated from several foreskins were pooled, collected by centrifugation (800 *g* for 1 min) and resuspended in Minimal Essential Medium containing antibiotics (penicillin 200 IU/ml, streptomycin 100 μg/ml, gentamicin 50 μg/ml). Cells were seeded in petri dishes and maintained at 37 °C in a humidified atmosphere of 6% CO_2 and

94% air (Davison *et al.*, 1980). In 1981 the same authors published a modification of this method and showed that the viability of HDMEC *in vitro* was extended when cultured in the presence of agents that elevated intracellular levels of cAMP (Davison & Karasek, 1981). Increased proliferation was in part attributable to improved cell attachment on seeding, which was approximately twice that seen in serum alone.

In 1980 Sherer and co-workers described their own method for the isolation and culture of HDMEC (Sherer *et al.*, 1980). Tissue was collected in buffered, antibiotic-containing K-Hanks' solution, washed (×5) in the same buffer with vortexing and, if not processed immediately, kept at 4 °C overnight. Collagenase and dispase were added and the samples incubated at 37 °C for 3 h. The contents of the flask were sieved twice through 153 μm nylon mesh followed by a 15 μm mesh. The remaining material was then collected and subjected to gentle centrifugation (1000 rpm) for 5 min at room temperature. Pellets were resuspended in 1 ml of growth medium (F-12K/FB5). Cells were seeded in F-12K medium supplemented with 5–20 % human or fetal bovine serum supplemented with 50 IU/ml penicillin and 50 μg/ml gentamicin. Antibiotics were omitted at the first change of medium after 2–4 days. Isolated cells were positive for factor VII-related antigen (von Willebrand factor) and had endothelial-specific Weibel–Palade bodies.

Since the early 1980s, different groups have attempted to improve the isolation and *in vitro* culture of HDMEC. Virtually all approaches have been based on the original description of Davison *et al.* (1980). In our experience a common problem is overgrowth by contaminating fibroblasts. Improved methods based on continuous gradient centrifugation of cells obtained from dermis after enzymatic and mechanical disaggregation of tissue fragments have been published by several groups (Marks *et al.*, 1985; Imcke *et al.*, 1989; Ruszczak *et al.*, 1990, Detmar *et al.*, 1990) and found to be easy and reproducible.

A similar method was recently published by Cornelius and co-workers (Cornelius *et al.*, 1995). This protocol is based on the work of Kubota and co-workers (1988) and of Ades and co-workers (1992), who have described an immortalised HDMEC line (HMEC-1) that retains many of the properties shown by primary isolates.

Media for the culture of human dermal microvascular endothelial cells

Microvascular endothelial cells have been difficult to culture for two reasons: (a) primary cultures are difficult to prepare; and (b) conventional cell culture media do not fully satisfy their growth requirements.

In 1987, Knedler and Ham introduced an optimised medium for the clonal growth of human microvascular endothelial cells that had minimal supplementation of serum (even as little as 0.7%). The medium contains hydrocortisone (1 μg/ml) and epidermal growth factor (10 ng/ml) (Knedler & Ham, 1987). Despite the fact that it is now known that Knedler and Ham were erroneously working with omental mesothelial and not endothelial isolates (see Scott and Bicknell, 1993) the medium (named MCDB131) has been of much use in the culture of true skin endothelial isolates. MCDB131 is characterised by a high magnesium concentration that has also been found to stimulate the growth of other endothelial isolates, such as that from decidua (Grimwood *et al.*, 1995).

The group of Karasek *et al.* have described the growth benefits of Eagle's Minimum Essential Medium (EMEM) containing different amounts of pooled human serum (10–50%) or maternal prepartum serum (up to 50%). EMEM may alternatively be replaced with Iscove's medium (Davison *et al.*, 1980; Davison & Karasek, 1981; Tuder *et al.*, 1987).

A procedure for the isolation and culture of HDMEC

Site of the skin biopsy

Best results are obtained with neonatal or children's foreskin (up to 9 years of age). Similar cell numbers and growth capacities are seen with young adult inguinal skin (patients up to 45–50 years old of either sex), the internal part of the upper arm or the upper leg. It has not proved possible to obtain a useful number of cells from biopsies taken in other body regions, notably from the trunk, buttock or abdomen. In older donors (over 60 years of age) successful isolates were seen only if skin biopsies were taken from an inguinal region (i.e. during routine saphenectomy). In 1986, Fuh and co-workers reported successful isolation of dermal microvascular endothelial cells from adult diabetic and non-diabetic individuals using skin from the abdominal region (Fuh *et al.*, 1986). Unfortunately, we have been unable to repeat the isolation.

Isolation protocol

Skin is collected in sterile Ca^{2+} and Mg^{2+}-free phosphate-buffered saline (PBS) or HEPES-buffered saline (HBS) (both purchased e.g. from GIBCO BRL, Life Technologies Ltd, Paisley, Scotland) containing 400 IU/ml penicillin, 400 IU/ml streptomycin and 0.5 μm/ml amphotericin B (GIBCO BRL, Paisley, Scotland). After 20–30 min samples are washed four times in sterile PBS or HBS supplemented with 100 IU/ml penicillin, 100 IU/ml streptomycin and 0.25 μg/ml amphotericin B (all purchased from GIBCO or PromoCell, Heidelberg, Germany) then cut into 5–6 mm^3 sections and processed immediately or stored in the same media at 4 °C for up to 24 h before use. The epidermis is removed from the dermis by immersion in 0.25% trypsin solution at 4 °C for 12–16 h (or overnight) followed by mechanical separation.

After removing the epidermis, dermal fragments are washed (×3) in sterile PBS or HBS. Trypsin activity is quenched by washing the skin samples with sterile PBS or HBS containing 10% fetal calf serum or bovine trypsin inhibitor at 37 °C (Sigma, St Louis, MO). Tissue fragments are then washed three times in sterile PBS or HBS supplemented with antibiotics and transferred to separation medium (medium 1: see Table 6.1) supplemented with antibiotics, magnesium sulphate and hydrocortisone.

After further washing in separation medium (medium 1), microvascular fragments are released into the medium by squeezing the papillary and subpapillary dermis using a no. 20 scalpel blade. Pressure is initially applied to the centre of each tissue section and the released fragments pushed outwards towards the periphery of the tissue and into the medium.

Endothelial cells have a different density to contaminants and can thus be separated by continuous density gradient centrifugation. Cells are filtered through a sterile 100 μm nylon mesh (Nitek, Tetko Inc., Elmsford, NY; Reichelt Chemie-Technik, Heidelberg, Germany) to remove debris. The endothelial cells are isolated by continuous Percoll (colloidal PVP-coated silicate of a density of 1.130 g/l, purchased from Sigma) gradient centrifugation. The Percoll gradient is prepared by centrifugation of 35% Percoll/PBS or Percoll/HBS solution at 30 000 *g* for 15 min. It is important to prepare the Percoll gradient carefully, and immediately before use. The Percoll stock solution is stored at 4 °C before use. The 35% Percoll/PBS or Percoll/HBS solution is prepared using cold (4 °C) PBS or HBS and mixed by gently shaking. Centrifugation should be at 4 °C in an ultracentrifuge with relatively high acceleration and a low braking rate (in our experiments the best results have been obtained with a Beckmann high-speed unit). The cell suspension, which after filtration through the nylon mesh contains both endothelial and non-endothelial cells, is washed (×3), resuspended in 1 ml of medium 1 or HBS and carefully placed onto the top of the gradient using a long pasteur pipette. Cells are centrifuged at 400 *g* for 15 min at 4 °C using Percoll density marker beads to standardise the gradient (Density marker beads for calibration of density gradient of Percoll: kit no. DMB-10, Sigma). The endothelial cells lie in the

Table 6.1. *Media for separation and culture of human dermal microvascular endothelial cells* in vitro

Medium 1. Transitory nutrition medium for separation of HDMEC
Iscove's medium
supplemented with: penicillin 100 IU/ml
streptomycin 100 μg/ml
amphotericin B 0.25 μg/ml
$MgSO_4$ 2.46 mg/ml (for a total amount of 2.46 g/l)
hydrocortisone 1 μg/ml

Medium 2. Seeding medium for HDMEC
Endothelial Cell Basal Medium (modified MCDB131, EBM)
(Clonetics, San Diego, CA; or PromoCell, Heidelberg, Germany)
supplemented with: penicillin 0.1mg/ml,
streptomycin 0.1 mg/ml
amphotericin B 0.25 μg/ml
hydrocortisone 1 μg/ml
epidermal growth factor 10 ng/ml
bovine pituitary extract 2 ml
fetal calf serum 10–20%
for adult endothelial cells additionally (Sigma): L-glutamine 1.46 mg/ml
magnesium sulphate 2.46 mg/ml
isobutylmethylxanthine 0.073 mg/ml (3.3×10^{-4} M)

Medium 3. Growth medium for HDMEC
Endothelial Cell Basal Medium (modified MCDB131, EBM)
(Clonetics, San Diego, CA; or PromoCell, Heidelberg, Germany)
supplemented with: penicillin 0.1 mg/ml,
streptomycin 0.1 mg/ml
amphotericin B 0.25 μg/ml
hydrocortisone 1 μg/ml
epidermal growth factor 10 ng/ml
fetal calf serum 2–10%

Medium 4. Growth medium for HDMEC (according to Cornelius et al., *1995)*
Endothelial Cell Basal Medium (modified MCDB131, EBM)
(Clonetics, San Diego, CA)
supplemented with: penicillin 0.1 mg/ml,
streptomycin 0.1 mg/ml
amphotericin B 0.25 μg/ml
hydrocortisone 1 μg/ml
epidermal growth factor 10 ng/ml
5×10^{-5} M dibutyryl cyclic AMP
2 mM glutamine
human serum 30%

gradient between 1.033 and 1.047 and may be removed using a sterile long glass pasteur pipette.

Cells are washed (×3) in medium 1, resuspended in 1 ml of medium 1 and transferred to the complete seeding medium (medium 2: see Table 6.1) based on endothelial cell basal medium (EBM; purchased from Clonetics, San Diego, CA, or from PromoCell, Heidelberg, Germany).

EBM from Clonetics and PromoCell is a ready-to-use, sterile-filtered liquid culture medium based on a MCDB131 formulated for cell growth in low serum. This formulation is optimised for an initial cell seeding density of approximately $5–10 \times 10^3$ cells/cm^2 and to permit growth to confluence (approximately 50 000–70 000 cells/cm^2). EBM produced by PromoCell is prepared from a set of nine different stock solutions and eight components added directly as solids. The pH is adjusted to 7.4 and osmolality to 275 ± 5 mosmol/kg. Both Clonetics and PromoCell products showed identical promotion of HDMEC growth.

After Percoll gradient centrifugation, 130 000–200 000 cells could be isolated from 1 cm^2 of human neonatal foreskin. If adult skin is used the number of the cells is significantly lower and depends on the localisation of the skin biopsy. The best results are obtained using human inguinal or ventral upper arm skin: 75 000–90 000 cells/cm^2.

Primary and long-term culture

To establish primary cultures, 35 mm or 80 mm plastic petri dishes (Falcon; Becton-Dickinson, Heidelberg, Germany) have been used. All plastic ware is coated with 10 μg/ml human fibronectin (GIBCO BRL, Paisley, Scotland). Primary isolates are seeded in 35 mm plates in a 3 ml volume at 20 000 cells/cm^2. For 80 mm plates 5 ml of cell/medium suspension of the same cell density is used. The first medium change is made after 3 days of culture. After washing twice with 37 °C PBS or HBS to remove unattached cells the medium is replaced by standard growth medium (medium 3). The growth medium is renewed every 2 days, and the cells passaged before they reach confluence.

Subcultivation is performed after washing subconfluent cells (×3) with PBS or HBS and exposure to 0.25% trypsin/PBS or trypsin/HBS solution for 10–20 min at 37 °C. The trypsin is quenched with medium 1 containing 30% fetal calf serum (3.0 ml for a 30 mm dish and 5.0 ml for an 80 mm dish, respectively), cells are collected in 50 ml plastic centrifuge tubes (BlueMax, Falcon; Becton-Dickinson, Heidelberg, Germany), resuspended and washed (×3) with medium 1 containing 30% serum. After washing, cells are resuspended in growth medium (medium 3: see Table 6.1.) and seeded at a density of 10 000 cells/cm^2.

Human dermal microvascular endothelial cells may be cultured for up to seven passages in a low-serum-containing medium. At the fourth passage the cells often lose their cobblestone morphology, becoming spindle-shaped and fibroblast-like. Addition of more than 10% FCS or human serum increases the life-span but changes cell morphology. Supplementation of growth medium with 20–30% FCS or 20% human serum and addition of cAMP increases the life-span to 14–16 passages. Use of pooled human serum at a concentration of 20–30% gives a similar extension of the life-span. Freshly isolated cells cultured in our standard medium (medium 3) are small and cuboidal, forming a cobblestone-like monolayer of cells. They show strong contact inhibition. After the fourth passage the cells become elongated, form bundles and show structures resembling capillaries with much overlap. Immunocytology has shown that the cells lose their endothelial markers, but remain positive for cytoskeletal proteins and mesenchymal markers. Serum-treated fresh isolates retain a cobblestone morphology for more than seven passages but are then fusiform and form a less regular monolayer of cells with overlapping processes and capillary-like structures. Serum-treated cells are larger and retain endothelial markers longer in culture. However, after eight to ten passages the cells also become fibroblastoid and lose their endothelial markers.

Possible mechanisms for the cessation of proliferation of HDMEC are: (a) cellular ageing, (b) terminal differentiation, or (c) inadequate culture conditions. We suggest that cellular ageing best describes the mechanism of cessation of proliferation. Thus, the cultured cells in standard medium undergo several key changes that characterise cellular ageing in other dermal cell populations such as fibroblasts: (1) their [^{3}H]thymidine labelling index remains high when the cells are young and gradually declines to zero at the end of their life-span. (2) The cells remains alive for several days at the end of the life-span. (3) The cells become enlarged at the end of the life-span. It is possible that some of the morphological changes of human dermal microvascular endothelial cells seen after the fourth passage are due to insufficient nutrition and/or the autocrine influence of endothelial cell-derived cytokines such as interferons that have been shown to induce a spindle-shaped morpholoy in HDMEC when added to the culture medium (Ruszczak *et al.*, 1990). For the above reasons we suggest the HDMEC only be used up to the second passage. We also recommend that isolation and culture be carried out in low-serum medium to avoid exposure to serum-derived exogenous growth factors and cytokines.

HDMEC may be cryopreserved using standard procedures. After removal with trypsin and washing (×3) in medium 1/30% serum, the cells are suspended in medium 1/10% dimethysulphoxide. Cells are placed in small (2 ml) cryotubes (Falcon, Becton-Dickinson) at 500 000 cells/ml. Slow freezing is achieved as follows: 30 min at 4 °C, 60 min at −20 °C, 60 min at −80 °C, followed by a final rapid freeze in liquid nitrogen. Frozen cells are viable in liquid nitrogen for up to 2 years.

If using a serum-free cell culture system, it is important to cryopreserve cells also in serum-free freezing medium (e.g. Cryo-SFM from PromoCell). When using serum-free medium, trypsin activity is quenched with bovine trypsin inhibitor, cells are frozen at $2–5\times10^6$ cells/ml in Cryo-SFM. (Caution: overexposure to trypsin is a common problem users encounter with serum-free cell culture systems.)

To thaw cryopreserved cells the vial is warmed to 37 °C and the cells immediately suspended in serum-free growth medium at a dilution of at least 1:10. After washing (×2), and collection at 200–220 *g* for 4 min the cells are gently resuspended at 10 000–20 000 cells/cm^2 in low-serum or serum-free growth medium for seeding.

Cell differentiation and immunocytochemical behaviour

The easiest method for the characterisation of freshly isolated or cultured cells is immune analysis of cytospin preparations. Cells are suspended in serum-free medium at a density of approximately 200 000 cells/ml and plated onto glass slides by centrifugation for 5 min in a Cytospin unit (Shandon, Runcorn, UK). After air drying, cells are fixed in acetone for 15 min at room temperature before incubation with antibodies. Slides may be stored dry at −20 °C until use.

Immunocytochemical characterisation of freshly isolated and cultured HDMEC is performed using the antibodies listed in Table 6.2. Visualisation is performed using either the APAAP or the avidin–biotin method. Labelling of Factor VIIIrAg is performed according to the Dako PAP kit (K 510), diaminobenzidine is used as a substrate and slides are counterstained with methyl green to enhance resolution.

HDMEC may also be stained with *Ulex europaeus* agglutinin 1 (UEA-1, Vector Laboratories, Burlingame, CA) using the avidin–biotin complex (ABC) method (Table 6.3).

Table 6.2. *Specificity of the monoclonal antibodies used for the characterisation of HDMEC*

Monoclonal antibody	Concentration	Specificity
BMA 120 (Behring, Germany)	1:50	Endothelial cells
Factor VIIIrAg (Dako, Denmark)	1:50	Endothelial cells
CK 8.13 (BioYeda, Israel)	1:75	Cytokeratins 1, 5–8, 10, 11, 18
PKK 2 (Labsystems, Finland)	1:100	Cytokeratins 7, 16, 17, 19
DE-R-11 (Dako, Denmark)	1:50	Intermediate filaments, desmin
V9 (Dako, Denmark)	1:500	Intermediate filaments, vimentin
Anti-Leu 6 (Becton-Dickinson, USA)	1:40	Leu 6
S 100 (Immunotech, Germany)	1:150	S100 protein
B1G6 (Immunotech, Germany)	1:50	β_2-Microglobulin
L 243 (Beckton-Dickinson, USA)	1:40	HLA-DR antigen

Table 6.3. *Immunocytochemical properties of freshly isolated cell suspensions, and primary and 2-week-old cultures*

	Proportion of labelled cells (%)		
Antigen	Freshly isolated	Primary cultures	Secondary culture
Ulex europaeus agglutinin 1	73	100	100
BMA 120	15	80	92
Factor VIII related antigen	80	95	98
CK 8.13	8	0	0
PKK 2	20	0	0
Desmin	0.6	0	0
Vimentin	54	100	100
Leu 6	0	0	0
S 100	0	0	0
HLA-Dr	20	0	0
β_2-Microglobulin	85	92	100

Preparation of cultures for electron microscopy

For ultrastructural examination, HDMEC cells are seeded onto Lux Thermanox tissue culture disks (Miles Laboratories, Naperville, IL) (for direct embedding) or into microtitre wells (for immunoelectron microscopy). For example, to perform immunoelectron microscopic examination cultures seeded on coverslips or in microtitre wells are washed (×2)

Table 6.4. *Step-by step cell separation protocol*

Step	Procedure	Time
1	Collection of skin samples in PBS or HBS supplemented with standard antibiotics (medium 1: see Table 6.1)	40 min
2	Separation of fat and deep dermal fragments (by cutting)	15 min
3	Washing four times (5 min each) in PBS or HBS supplemented with standard antibiotics (medium 1)	20 min
4	Transfer into a new PBS or HBS solution and cut into 5–6 mm^3 (max. 2×3 mm) tissue fragments	15–20 min
5	Transfer into 0.25% trypsin/PBS or 0.25% trypsin/HBS solution for epidermis separation	12–18 h at 4 °C; 60 min at 37 °C
6	Epidermis separation, transfer of de-epidermised dermal fragments into 10% serum/PBS or 10% serum/HBS solution at 37 °C	10 min
7	Washing of de-epidermised dermal fragments three times (5 min each) in sterile PBS or HBS (20 °C) supplemented with standard antibiotics	15 min
8	Transfer of tissue fragments into the medium 1 for further cell separation	10 min
9	Gentle short washing in medium 1	3 min
10	Gentle scraping of tissue fragments using no. 20 scalpel blade	20–30 min
11	Removal of large tissue from the above	10 min
12	Filtration of the cell suspension through 100 μm nylon mesh to separate debris	15 min
13	Washing cell suspension twice in 10 ml of medium 1 by centrifugation (200 *g*, 4 °C)	2×10 min
14	Resuspension in 10 ml of medium 1	
15	Preparation of Percoll gradient (35% Percoll/PBS or HBS) by centrifugation at 30 000 *g*, 4 °C	15 min
16	Washing of resuspended cells (see step 14) by centrifugation at 200 *g* at 4 °C, and resuspension in 1 ml of medium 1	5 min
17	Seeding of cell suspension (see step 16) on the top of the freshly prepared Percoll gradient	2–3 min
18	Centrifugation at 400 *g* at 4 °C using marker beads	15 min
19	Removal of endothelial cell fraction (1.033–1.047)	2–3 min
20	Resuspension in 5 ml of medium 1 and washing three times in 5 ml of medium 1; centrifugation at 200 *g* for 10 min	3×10 min 1×10 min
21	Preparation of human fibronectin coated plastic plates	25–30 min
22	Resuspension in medium 2 (see Table 6.1), cell counting, plating at a density of 10 000–20 0000 cells/ml	30 min

in PBS, pre-fixed for 15 min at room temperature with 0.1% glutaraldehyde in PBS and incubated with antibody for 30 min at 37 °C. This is followed by a second 30 min incubation at 37 °C with colloidal gold-conjugated IgG and then processing for electron microscopy.

Closing remarks

Using the protocols described to isolate and culture endothelial cells from skin provides an appropriate and physiologically relevant system to evaluate skin capillary function. Both normal and pathological endothelium may be analysed. Examples of the latter include the isolation and characterisation of skin microvascular endothelium from diabetic patients (Fuh *et al.*, 1986), human adult haemangiomas (Tuder *et al.*, 1987), connective tissue diseases (Kahaleh, 1990), as well as cell populations derived from Kaposi's sarcoma (Ruszczak *et al.*, 1991; Young *et al.*, 1994). Unique biological properties of skin capillary endothelium have recently been shown in response to factors influencing growth, differentiation, antigen presentation and migration.

The co-cultivation of HDMEC with other cell populations obtained from the skin, such as fibroblasts, dendritic cells, epidermal or hair-derived keratinocytes, with or without a collagen matrix as an *in vitro* dermis substitute, offers new possibilities to study cell–cell interactions and processes controlling the normal and abnormal function of skin.

References

Ades, E.W., Candal, F.J., Swerlick, R.A., George, V.G., Summers, S., Bosse, D.C. & Lawley, T.J. (1992). HMEC-1: establishment of an immortalised human microvascular endothelial cell line. *J. Invest. Dermatol.*, **99**, 683–90.

Cornelius, L.A., Nehring, L.C., Roby, J.D., Parks, W.C. & Welgus, H.G. (1995). Human dermal microvascular endothelial cells produce matrix metalloproteins in response to angiogenic factors and migration. *J. Invest. Dermatol.*, **105**, 170–6.

Davison, P.M. & Karasek, M.A. (1978). Factors affecting the growth and morphology of rabbit skin endothelial cell *in vitro*. *Clin. Res.*, **26**, 208A.

(1981). Human dermal microvascular endothelial cells *in vitro*: effect of cyclic AMP on cellular morphology and proliferation rate. *J. Cell. Physiol.*, **106**, 253–8.

Davison, P.M., Bensch, K. & Karasek, M.A. (1980). Isolation and growth of endothelial cells from the microvessels of the newborn human foreskin and cell culture. *J. Invest. Dermatol.*, **75**, 316–21.

Davison, P.M., Bensch, K. & Karasek, M.A. (1983). Isolation and long-term serial cultivation of endothelial cells from the microvessels of the adult human dermis. *In Vitro*, **19**, 937–45.

Detmar, M., Ruszczak, Z., Imcke E. & Orfanos C.E. (1990). Effects of tumor necrosis factor-alpha (TNFα) on cultured microvascular endothelial cells derived from human dermis. *J. Invest. Dermatol.*, **95**, 219–22.

Fuh, G.M., Bensch, K., Karasek, M.A. & Kramer, R.H. (1986). Synthesis of basement membrane-specific macromolecules by cultured human microvascular endothelial cells isolated from skin of diabetic and nondiabetic subjects. *Microvasc. Res.*, **32**, 359–70.

Grimwood, J., Bicknell, R. & Rees, M.C.P. (1995). The isolation, characterization and culture of human decidual endothelium. *Hum. Reprod.*, **10**, 101–8.

Imcke, E., Ruszczak, Z., Mayer-da-Silva, A. & Orfanos, C.E. (1989). *In vitro* cultivation of dermal microvascular endothelial cells and their immunocytochemical and electron microscopic characterisation. *J. Invest. Dermatol.*, **92**, 499(A).

Kahaleh, M.B. (1990). The role of vascular endothelium in the pathogenesis of connective tissue diseases: endothelial injury, activation, participation and response. *Clin. Exp. Rheumatol.*, **8**, 595–601.

Knedler, A. & Ham, R.G. (1987). Optimized medium for clonal growth of human microvascular endothelial cells with minimal serum. *In Vitro Cell. Dev. Biol.*, **23**, 481–91.

Kubota, Y., Kleinman, H.K., Martin, G.R. & Lawley, T.J. (1988). Role of laminin and basement membrane in the morphological differentiation of human endothelial cells into capillary-like structures. *J. Cell Biol.*, **107**, 1589–98.

Marks, R.M., Czerniecki, M. & Penny, P. (1985). Human dermal microvascular endothelial cells: an improved method for tissue culture and a description of some singular properties in culture. *In Vitro*, **21**, 627–35.

Ruszczak, Z., Detmar, M., Imcke, E. & Orfanos, C.E. (1990). Effects of rIFN-alpha, -beta, and -gamma on the morphology, proliferation, and cell surface antigen expression of human dermal microvascular endothelial cells *in vitro*. *J. Invest. Dermatol.*, **95**, 693–9.

Ruszczak, Z., Detmar, M., Stadler, R., Bratzke, B. & Orfanos, C.E. (1990). rIFN-α inhibits the proliferation and rIFN-γ upregulates cell surface antigen expression of HIV-associated Kaposi's sarcoma-derived cells *in vitro*. *Arch. Dermatol. Res.*, **283**, 49–50.

Scott, P.A.E. & Bicknell, R. (1993). The isolation and culture of microvascular endothelium. *J. Cell Sci.*, **105**, 269–73.

Sherer, G.K., Fitzharris, T.P., Faulk, W.P. & Le Roy, E.C. (1980). Cultivation of microvascular endothelial cells from human preputial skin. *In Vitro*, **16**, 675–84.

Tuder, R.M., Young, R. Karasek, M.A. & Bensch, K. (1987). Adult cutaneous hemangiomas are composed of nonreplicated endothelial cells. *J. Invest. Dermatol.*, **89**, 594–7.

Young, J., Xu-Y., Zhu-C., Hagan, M.K., Lawley, T. & Offermann, M.K. (1994). Regulation of adhesion molecule expression in Kaposi's sarcoma cells. *J. Immunol.*, **152**, 361–73.

7

Microvascular endothelium from adipose tissue

Stuart K. Williams

Introduction

The pioneering work of Lewis in the 1920s provided the groundwork for the eventual establishment of both large blood vessel and microvessel derived endothelial cell cultures (Lewis, 1922). The earliest report of methods for the isolation of microvessel derived endothelial cells, and in particular endothelial cells derived from microvascularised fat, was a publication by Wagner and colleagues in 1972. The choice of fat as a source of tissue was based on its ready availability, its high density of microvascular endothelial cells and previously established methods for the digestion of adipose tissue using proteolytic enzyme solutions containing crude clostridial collagenase. Thus the choice of fat tissue as a source of endothelium was not based on specific properties to be studied. In fact Wagner's report provided initial evidence that not only microvascular endothelium but endothelium in general exhibited specific metabolic activities. The tenet that endothelium existed as a metabolically inert 'Cellophane' lining was initially questioned by Wagner's evaluation of isolated fat-derived microvascular endothelial cells.

Since these early studies using adipose tissue derived endothelium, investigators have begun to evaluate microvascular endothelial cells from essentially every vascularised tissue in the body. Continued interest in adipose tissue endothelial cells persists for the following reasons: (1) studies of obesity and general nutrition, (2) studies of the augmentation of wound healing by the presence of adipose tissue and (3) the use of adipose tissue derived endothelium for cell transplantation. Obesity and nutrition studies have utilised adipose-derived endothelium to evaluate their role in anabolic and catabolic activity and in the regulation of nutrient and waste product transport between blood and adipocytes. Investigators evaluating wound healing have taken an interest in adipose endothelium due to the omnipresence of adipose tissue in the area of soft tissue wounds and the fact that numerous biomedical implants are placed subcutaneously or intraperitoneally within adipose tissue. Finally, researchers have developed methods for the transplantation of endothelial cells onto biomedical devices using adipose tissue derived endothelium as a source of cells. This last research application has been actively studied over the past 15 years and the field has recently been reviewed (Williams, 1995).

Characterisation

The microvasculature of adipose tissue is characterised by the presence of arteriolar, venular and capillary endothelium. With the exception of perirenal adipose tissue the predominant microvascular element is capillary endothelium, characterised as continuous capillaries by the presence of tight junctions, lack of fenestrae or junctional discontinuities. Inter-endothelial cell junctions exhibit cadherins, PECAM and ZO-1 proteins typical of junctional complexes which regulate solute and cellular exchange. Adipose microvasculature and specifically capillary endothelium exhibits very active solute transcytosis, utilising micropinocytic vesicles which are present at relatively high densities (Wagner & Matthews, 1975). Luminal diameters of these capillaries are relatively constant, at an average of 6 μm. Adipose capillaries are metabolically extremely active and exhibit numerous enzymatic activities and hormone receptors. Endothelial cell specific lectins (e.g. UEA-1, *Bandeiraea simplicifolia*, depending on species) react actively with adipose endothelial cells. On the other hand, staining of intact adipose tissue with antibodies specific for von Willebrand factor (vWF) results in active staining of only venular endothelial cells while adipose tissue capillary endothelium exhibits variable reactivity for this antigen. Electron microscopic evaluation of adipose capillary endothelium provides evidence of the limited presence of Weibel–Palade bodies, explaining the poor staining with antibodies to vWF.

Source of cells

Although adipose tissue is nearly ubiquitous in all bodily tissues, differences exist in the the cellular composition of adipose tissue between species and within a single species. Experience in obtaining adipose tissue from these species and subsequent isolation and culture of endothelial cells has established optimal sites of adipose tissue procurement. Each of these sites is now described for each animal species which have been evaluated in the author's laboratories.

Human

The original source of human adipose tissue used for subsequent microvessel endothelial cell isolation was omental fat deposits associated with omentum (Jarrell *et al.*, 1984). Investigators have also reported the isolation of endothelial cells from perirenal fat and breast fat. The use of subcutaneous fat removed during laparotomy has also been reported (Williams *et al.*, 1989). The author's laboratory has also evaluated the use of abdominal wall subcutaneous fat procured from humans through the use of liposuction (Williams *et al.*, 1994*b*). Liposuction has emerged as the most easily obtained source of fat as the procedure is relatively non-invasive requiring a small skin incision (or alternatively the use of a trocar) and the insertion of a relatively small liposuction cannula attached to a 60 ml syringe. Suction is created by pulling back on the syringe plunger and the adipose tissue is repeatedly probed with the cannula, resulting in the removal of 50 ml of adipose tissue in 5 min or less. Another advantage of this technique is the disruption of the tissue into small pieces (approximately 1–3 mm^3) which need no further dissection before collagenase digestion. The sources of human fat for endothelial cell isolation have been extensively characterised and subcutaneous fat removed by liposuction contains predominantly endothelial cells and adipocytes (Williams *et al.*, 1994b). Humans are the only species with suitable subcutaneous fat deposits for endothelial cell isolation.

Rat

The epididymal fat pad of adult rats was the first source of fat reported for the isolation of microvessel endothelial cells (Wagner *et al.*, 1972; Wagner & Matthews, 1975). This source remains the most commonly used site for fat removal from the rat. Female rats have also been used for adipose tissue retrieval as they contain a significant amount of localised fat in the intraperitoneal space around the uterine horns. Caution must be exercised when isolating this fat as both the epididymal fat pads and uterine fat pads have a serosal lining composed of a single layer of mesothelial cells. This layer of mesothelium can serve as a source of contaminating mesothelial cells in the final isolates. Scraping the fat pads after their removal will remove a significant number of these cells.

Mouse

Fat located in the intraperitoneal cavity is the major source of fat from the mouse (Jemison *et al.*, 1993). Successful isolation of fat from the mouse is dependent upon the nutritional status and age of the animals. Younger animals, tumour-bearing animals and SCID mice have limited fat deposits: however, careful dissection and evaluation of the peritoneal cavity under a dissecting microscope will reveal fat tissue suitable for endothelial cell isolation. We have successfully isolated and cultured endothelial cells from SCID mouse uterine horn and epididymal fat pads.

Dog

Most canine species have extensive fat deposits in the intraperitoneal space, but, as observed with all species except humans, the dog has minimal subcutaneous fat. The most easily accessible fat source in the dog is a rich fat deposit located in association with the inguinal ligament (Williams *et al.*, 1994a). This fat is procured from anaesthetised dogs by making a midline incision between the umbilicus and the diaphragm notch. The incision is continued through the skin and muscle until the serosal surface of the peritoneal space is observed. Clamps are placed on this serosal layer and the layer transected using cautery. The underlying fat exposed is secured proximally and distally with large haemostats, the fat tied and localised with 4-0 silk proximal to each haemostat, and fat between the haemostats removed using a scalpel. The haemostats are removed and the incision closed in three layers. Characterisation of this fat has identified adipocytes and endothelial cells as the predominant cell types.

Pig

Although the pig contains huge quantities of fat, characterisation of different fat deposits in this animal has revealed only one anatomical site with fat suitable for endothelial cell isolation (Young *et al.*, 1992). This site is located along the abdominal wall and is identified as properitoneal fat. It is reached only by laparotomy of anaesthetised animals. The serosal layer covering the fat is transected and pulled away as a sheet. The fat deposits occur in patches and are removed using fine dissection.

Methods of isolation

Equipment
Dubnoff shaking water bath
50 ml Nalgene Erlenmeyer flasks
Small magnetic stir bars
Table-top centrifuge

Solutions and supplies

Crude clostridial collagenase. A significant difference exists between lots of collagenase with respect to digestion efficiency and cell function. Collagenase lot selection criteria for adipose tissue digestion have been reported (Williams *et al.*, 1995)

Dulbecco's cation-free phosphate-buffered saline (DCF-PBS, pH 7.4) containing 0.1% essentially fatty acid free bovine serum albumin

Culture Medium 199E containing 15% fetal bovine serum, endothelial cell growth supplement (ECGS) and heparin (Thornton *et al.*, 1983). The ECGS is prepared from bovine hypothalamus and a bioassay performed to determine optimal concentration. Heparin lots also show significant variability in performance. We routinely perform bioassays on heparin to determine optimal concentration

Gelatin coated tissue culture plastic. Tissue culture plates are treated with 1% gelatin

The methods for endothelial cell isolation from adipose tissue isolated from all species are essentially identical and thus one method will be described that is applicable to all species and fat sources. The only differences between the fat sources during isolation are the amount of tissue dissociation needed prior to collagenase digestion and the concentration of collagenase used.

Fat obtained from humans using liposuction is used without further mincing, while fat from all other anatomical sites in humans and from all other species is minced using fine scissors. The minced tissue is washed once with DCF-PBS (pH 7.4). The fat is then placed in 50 ml Erlenmeyer flasks which contain crude clostridial collagenase, an equal concentration (mg/ml) of bovine serum albumin, DCF-PBS and a magnetic spin bar. Each flask contains 10 ml of collagenase/albumin solution and 10 ml of fat tissue. The type and concentration of collagenase used for adipose tissue digestion is highly dependent on the lot and the manufacturer. The choice and use of collagenase for tissue digestion have recently been evaluated and reported (Williams *et al.*, 1995). Review of the author's experience with numerous lots of collagenase indicates that 2–4 mg/ml concentrations of crude clostridial collagenase are optimal for all species except the dog, for which more concentrated collagenase solutions are generally necessary to achieve complete digestion.

The Erlenmeyer flask containing the fat/collagenase slurry is placed in a Dubnoff shaking water bath at 37 °C and agitated vigorously. The efficiency of collagenase digestion can be evaluated temporally by withdrawing samples of digesting tissue at regular intervals. Digestion is complete when single free adipocytes are observed with minimal clumps, and endothelial cells occur as predominantly single cells with occasional multicellular tubes. This digestion normally takes 30 min.

The cellular slurry is subsequently placed in a plastic (polypropylene) 15 ml centrifuge tube and centrifuged (850 *g*, 5 min), resulting in a pellet of vascular elements and a floating cake of adipocytes and oil. The pellet is suspended in DCF-PBS and centrifuged (450 *g*, 4 min). This final pellet is suspended in a medium appropriate for either use of the cells in an immediate assay, cell transplantation or culture. The methods described below are for the establishment of cells in culture.

The pellet of vascular elements from the collagenase digestion is suspended in tissue culture medium as defined above. Cells are plated onto tissue culture plastic previously treated with 1% gelatin. This primary isolate contains endothelium as well as red blood cells. The culture dishes are placed in a 37 °C, 5% CO_2 incubator for 20 min to allow endothelial cell attachment. Again correct lot selection of collagenase is critical, since certain commercially available collagenases support rapid tissue digestion but result in cells with very poor adherence capacity on gelatin coated plastic. Following 20 min of incubation the culture surface is washed vigorously with tissue culture medium. This washing is performed with a pasteur pipette and the entire culture surface should be subjected to the fluid shear stress created when medium is evacuated from the pipette. Previous studies have confirmed that primary isolates of fat-derived microvascular endothelial cells exhibit a shear-resistant attachment to plastic surface as early as 5 min after plating. Other contaminating cell types such as pericytes, fibroblasts and smooth muscle cells require a significantly longer period to establish shear-resistant attachment. Under controlled shear stress conditions microvessel endothelial cells remain adherent to shears in excess of 75 dynes/cm^2. Following this washing the remaining adherent cells are placed in a 37 °C, 5% CO_2 incubator. These cultures are fed again after 24 h of primary culture. The morphological phenotype of cell isolates after 24 h of culture is illustrated in Fig. 7.1*a*.

Establishment and maintenance of cultures

Primary isolates established as described above are fed three times weekly. Following 24 h of culture fat-derived microvessel endothelial cells begin to exhibit numerous morphological phenotypes ranging from single cobblestone morphologies, through diamond ring morphologies to cells exhibiting numerous filopodia and cytoplasmic extensions. These cultures are morphologically different from large vessel derived endothelial cell cultures after 24 h in culture. This morphological differentiation continues in subsequent passages.

As adipose microvascular endothelial cells begin to approach confluence in culture their morphology begins to change from a more stellate, fibroblastic phenotype to a more ordered oblong morphology (Fig. 7.1*b*). At confluence (Fig. 7.1*c*) the cells are tightly packed and show the monolayer phenotype exhibited by large vessel endothelial cells. Primary cultures of these cells are routinely analysed prior to first passage for markers considered to be highly characteristic of endothelial cells. These markers include the expression of vWF, receptors of acetylated low density lipoprotein (Fig. 7.1*d*), PECAM and specific lectins. In addition, endothelial cells exhibit the ability to undergo vasculogenesis, or tube formation, when placed on basement membrane collagen. Fig. 7.1*e* illustrates the morphological change in a culture of adipose tissue derived microvascular endothelial cells plated onto Matrigel reconstituted basement membrane extracellular matrix.

During the first few days of primary culture investigators can selectively clone cells on the basis of their morphological phenotype. This process is known as selective weeding and is accomplished using an inverted phase contrast microscope. The microscope is placed inside a laminar flow hood and primary cultures of fat microvessel endothelial cells are visualised and removed within cloning cylinders. This morphological cloning results in cell lines with relatively stable yet distinct morphologies. The anatomical origins of these different morphologies are presently unknown. If the cultures are not selectively cloned but are maintained as a morphologically heterogeneous population of cells, the cell types with the highest proliferative capacity will predominate in the culture at confluence.

Long-term culture and passage

Fat-derived microvessel endothelial cells are grown to confluence and passaged at split ratios between 1:4 and 1:10 using trypsin/EDTA to release cells. Gelatin coated tissue culture plastic is used during subsequent passages. The relative life-span of adipose tissue microvessel endothelial cells is species dependent, with human fat microvessel endothelium exhibiting comparatively more limited cumulative population doublings (CPD) before senescence. Under the culture conditions described, human fat microvessel endothelial cells are capable of 30–35 CPD before senescence while all other species exhibit >40 CPDs prior to senescence. This CPD level is lower than

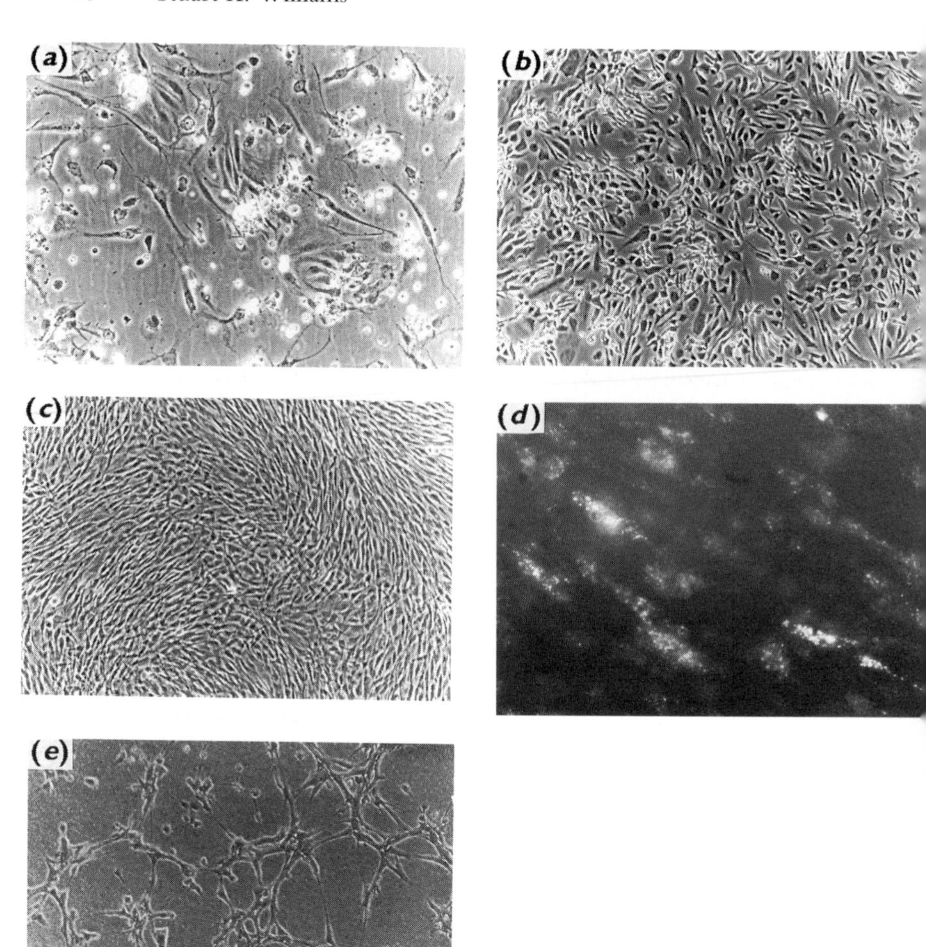

Fig. 7.1. Light microscopy illustrating the morphological phenotypes of adipose tissue derived microvascular endothelial cells during different stages of culture. All micrographs are phase contrast except (e) which is an epifluorescence image. (*a*) Primary isolate following 24 h of culture. (*b*) Primary isolate following 4 days of culture. (*c*) Primary isolate at confluence. (*d*) DiI-acetylated LDL staining of a primary isolate of cells at confluence. (e) Primary isolate grown to confluence and plated onto Matrigel basement membrane extracellular matrix.

the 65–75 CPD level observed for human large vessel derived endothelial cells. This suggests either the optimal growth conditions for human microvessel endothelial cells have not yet been obtained or these cells exhibit a programmed early senescence in culture.

Cryopreservation

Equipment

−20 °C freezer
−70 °C freezer
Liquid nitrogen storage tank
Table-top centrifuge

Solutions and supplies

Dimethylsulphoxide (DMSO)
Tissue culture medium (as described above)
Cryovials (2 ml)

Prepare DMSO solution by mixing 9 parts tissue culture medium with 1 part DMSO. Place this solution on ice. Microvascular endothelial cells are grown to confluence in a T-75 flask and dissociated using a rubber policeman. The cellular supernatant is placed in a 15 ml conical centrifuge tube and cells pelleted by centrifugation at 100 *g* for 4 min. Aspirate the supernatant and add 1 ml of the DMSO/media solution. Suspend the pelleted cells by light aspiration and transfer the suspended cells to a cryovial. Place vials on ice for 30 min then transfer the vial to a Styrofoam vial rack. Place the rack with vial in a −20 °C freezer for 2 h, then transfer the rack to a −70 °C freezer for 12 h. The vial is then transferred to a liquid nitrogen storage tank. Vials are thawed by quick immersion in a 37 °C water bath with constant agitation. Transfer the cells and media to a 15 ml centrifuge tube and immediately add 10 ml of tissue culture medium. Pellet cells by centrifugation (100 *g*; 4 min) and suspend cells in 5 ml of tissue culture medium. Seed cells onto gelatin coated tissue culture plastic and bring the volume of medium to a level appropriate for the flask used.

References

Jarrell, B.E., Levine, E.M., Shapiro, S.S., Williams, S.K., Carabasi, R.A., Mueller, S.N. & Thornton, S.C. (1984). Human adult endothelial cell growth in culture. *J. Vasc. Surg.*, **1**, 757–64.

Jemison, L.M., Williams, S.K., Lublin, F.D., Knobler, R.L. & Korngold, R. (1993). Interferon-inducible endothelial cell class II major histocompatibility complex expression correlates with strain- and site-specific susceptibility to experimental allergic encephalomyelitis. *J. Neuroimmunol.*, **47**, 15–22.

Lewis, W.H. (1922). Endothelium in tissue culture. *Am. J. Anat.*, **30**, 39–59.

Thornton, S.C., Mueller, S.N. & Levine, E.M. (1983). Human endothelial cells: cloning and long term serial cultivation employing heparin. *Science*, **222**, 623–4.

Wagner, R.C. & Matthews, M.A. (1975). The isolation and culture of capillary endothelium from epididymal fat. *Microvasc. Res.*, **10**, 286-97.

Wagner, R.C., Kreiner, P., Barnett, R.J. & Bitensky, M.W. (1972). Biochemical characterization and cytochemical localization of a catecholamine-sensitive adenylate cyclase in isolated capillary endothelium. *Proc. Natl. Acad. Sci. USA*, **69**, 3175–9.

Williams, S.K. (1995). Endothelial cell transplantation. *Cell Transplant.*, **4**, 401–10.

Williams, S.K., Jarrell, B.E., Rose, D.G., Pontell, J., Kapelan, B.A., Park, P.K. & Carter, T.L. (1989). Human microvessel endothelial cell isolation and vascular graft sodding in the operating room. *Ann. Vasc. Surg.*, **3**, 146–52.

Williams, S.K., Rose, D.G. & Jarrell, B.E. (1994*a*). Microvascular endothelial cell sodding of ePTFE vascular grafts: improved patency and stability of the cellular lining. *J. Biomed. Mater. Res.*, **28**, 203–12.

Williams, S.K., Wang, T.F., Castrillo, R. & Jarrell, B.E. (1994*b*). Liposuction derived human fat used for vascular graft sodding contains endothelial cells and not mesothelial cells as the major cell type. *J. Vasc. Surg.*, **19**, 916–23.

Williams, S.K., McKenney, S. & Jarrell, B.E. (1995). Collagenase lot selection and purification for adipose tissue digestion. *Cell Transplant.*, **4**, 281–9.

Young, C., Jarrell, B.E., Hoying, J.B. & Williams, S.K. (1992). A porcine model for adipose tissue-derived endothelial cell transplantation. *Cell Transplant.*, **1**, 293–8.

8

Endothelium of the female reproductive system

Yuan Zhao and Margaret C. P. Rees

Introduction

The vasculature of the female reproductive tract is intimately involved in the processes of ovulation, menstruation, implantation, placentation and fetal development. These tissue beds have the unique property of undergoing benign angiogenesis, a process otherwise restricted to tissue repair. This chapter will describe the isolation and culture of the endothelium of the corpus luteum and of human first trimester (early pregnancy) decidua. It will also briefly cover the endothelium of other female reproductive tissues. No description will be given of the isolation and culture of human umbilical vein endothelium (HUVEC) since this has been described in Chapter 3.

The ovary

Mammalian ovarian follicles possess a well-developed capillary bed that delivers nutrients, growth factors and gonadotrophins required for the growth and maturation of the follicle and subsequent transformation into a corpus luteum. Ovarian follicles are thought to control the development of their own capillary bed, although there is no strong evidence to support this. Various regulators of angiogenesis have been identified in ovarian tissue, including epidermal growth factor (EGF), transforming growth factor α (TGFα), transforming growth factor β (TGFβ), basic fibroblast growth factor (bFGF), tumor necrosis factor α (TNFα), interleukin-1 (IL-1), inhibin and activin. The precise role played by each factor is not understood. Inhibition of ovarian angiogenesis could give rise to the development of new contraceptive agents.

During the menstrual and oestrous cycles, the ovarian follicle ruptures at ovulation and then develops into the corpus luteum which contains a well-developed microvasculature. Attempts have been made to isolate and culture endothelial cells from bovine and rabbit corpora lutea (Spanel-Borowski & Bosch, 1990; Bagavandoss & Wilks, 1991; Fenyves *et al.*, 1993) and detailed methods are given in the following section. There are no reports to date of the isolation and culture of the endothelium of the human corpus luteum.

Characterisation of the endothelial cells of corpora lutea

Several different phenotypes of endothelial cells have been characterised in the cultured endothelium from bovine corpus luteum, which illustrates the morphological heterogeneity of this particular microvascular bed. Spanel-Borowski & Bosch (1990) distingushed four different phenotypes (Fig. 8.1). They described subconfluent type 1 cells as being 'arranged in strands or groups of uniform-sized cells that displayed a central core of a non-transparent endoplasm surrounded by a broad rim of transparent ectoplasm'. The cells were isomorphic and of cobblestone morphology when they reach confluence (Fig. 8.1*a*, *b*). In contrast, type 2 cells were crescent-shaped and polymorphic at confluence (Fig. 8.1*c*, *d*). Type 3 and type 4 cells both exhibited veil-like and harpoon-like cell processes (Fig. 8.1*e*), but differed in morphology at confluence. Thus type 3 cells were spindle-shaped (Fig. 8.1*f*), but type 4 cells showed web-like filaments on the cell surface (Fig. 8.1*g*). Scanning electron microscopy revealed surface differences between the four cell types in terms of both the density of surface projections and the cell border (Spanel-Borowski & Bosch, 1990; Spanel-Borowski, 1991).

The endothelium of bovine corpora lutea is positive for several adhesion molecules, including neuronal cell adhesion molecule (NCAM), N-cadherin (N-CAD) and E-cadherin (E-CAD), although the four phenotypes again differ (Spanel-Borowski & Fenyves, 1994). TNFα receptors have been detected on endothelial cells of pig corpora lutea (Richards & Almond, 1994). TNF, IL-1α and IL-1β have been shown markedly to inhibit proliferation of cultured endothelial cells from rabbit corpus luteum (Bagavandoss & Wilks, 1991).

Isolation and culture of the endothelial cells of corpora lutea

Corpora lutea are excised from ovaries and cut into 1 mm^2 pieces. The fine tissue fragments are further dissociated by vigorous pipetting. The resulting cell suspension is serially sieved through 150 and 72 μm^2 pore size meshes to remove cell aggregates. The dislodged cells are then subjected to density centrifugation in 50% Percoll to separate endothelial cells from red blood

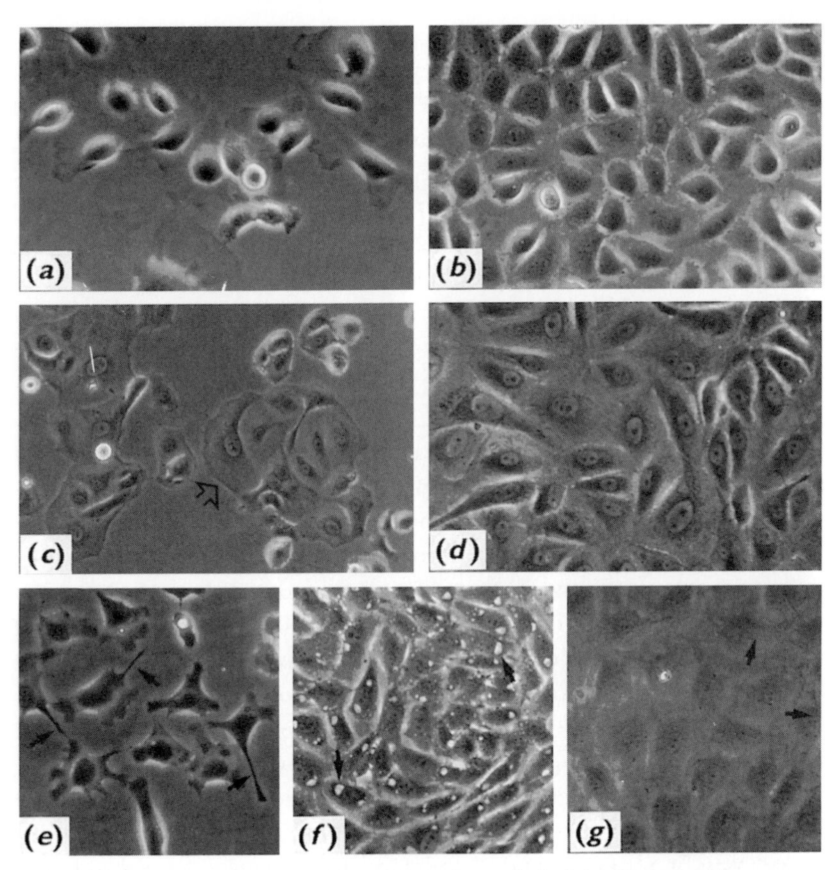

Fig. 8.1. Differential morphology of microvascular endothelial cells derived from bovine corpus luteum. (*a*), (*b*) Type 1. (*a*) Subconfluent culture. First passage. ×340. (*b*) Confluent isomorphic culture of cobblestone morphology. ×280. (*c*), (*d*) Type 2. (*c*) Subconfluent culture with a cell island bordered by a crescent-shaped cell (arrow). First passage. ×130. (*d*) Confluent polymorphic culture. ×280. (e)–(*g*) Types 3 and 4. (e) Subconfluent culture of single cells with veil-like and harpoon-like cell processes (arrows). First passage. Type 3 and 4 look similar. ×340. (*f*) Confluent culture of type 3. Distinct vacuoles are seen in spindle-shaped cells (arrows). ×280. (*g*) Culture of type 4 confluent for 3 weeks. Delicate web-like filaments appear on the surface of round cells (arrows). ×280.

cells and luteal cells. Enriched endothelial cells are recovered in the fraction at the bottom of the centrifuge tubes and washed three times with DF medium (1:1 Dulbecco's Modified Eagle Medium:Ham's F-12 medium) to remove Percoll. Cells are then seeded onto collagen coated plates. Unattached cells are removed by changing the medium after a 12 h incubation. Endothelial cells are further purified by removal of endothelial colonies from a confluent culture with a bent eppendorf tip. The endothelial colonies are disaggregated by treatment with trypsin/EDTA before seeding. The pure endothelial cells are passaged 1:4. DF medium supplemented with 5% fetal calf serum (FCS) and endothelial mitogen (10 mg/ml; Paesel, Frankfurt, Germany) are optimal for the culture of bovine corpus luteum endothelial cells (Spanel-Borowski & Bosch, 1990). In contrast, endothelial basal medium (EBM) supplemented with 10% FCS and bFGF (2 ng/ml) is good for the culture of rabbit corpus luteum endothelial cells (Bagavandoss & Wilks, 1991).

The endometrium and decidua

The endometrium is a mucosa served by a microvascular blood supply that differentiates cyclically under the influence of sequential oestradiol and progesterone secreted by the ovary during each menstrual cycle. Regeneration of the microvasculature is initiated from the remaining post-menstrual vascular stumps in the basalis of the endometrium and includes well-differentiated coiled arterioles, venules and capillaries (Fig. 8.2) (Kaiserman-Abramof & Padykula, 1989). During menstruation, vasoconstriction with shedding of some endometrium occurs. Since the endothelium lacks the steroid receptors present in the epithelial and stromal components, it has been proposed that angiogenic factors produced by the latter two cell types control the process of endometrial angiogenesis. Various angiogenic factors, i.e. acidic and basic fibroblast growth factor (aFGF, bFGF), vascular endothelial growth factor (VEGF), pleiotrophin (PT), midkine (MK), platelet derived endothelial cell growth factor (PDECGF) and transforming growth factor β1 (TGFβ1) have been detected in endometrium or cultures of epithelium and stroma. Expression of VEGF and MK are regulated by physiological levels of oestradiol and progesterone in endometrial epithelial isolates (Zhang *et al.*, 1995).

During pregnancy, endometrium undergoes extensive changes to become the tissue known as decidua. This conversion is a key process in successful implantation and pregnancy. Decidualisation involves the formation of enlarged stromal cells, a reduction in epithelial tissue and an increase in vascular density. Further crucial events in early pregnancy are the development of the maternal placenta and the migration of trophoblastic cells into the decidua. Trophoblastic invasion occurs in two waves, one in the first and one in the second trimester. The trophoblasts destroy vascular smooth muscle and the end result is that the muscular walled vascular spiral arterioles of the endometrium are converted into a low-resistance vascular bed which can

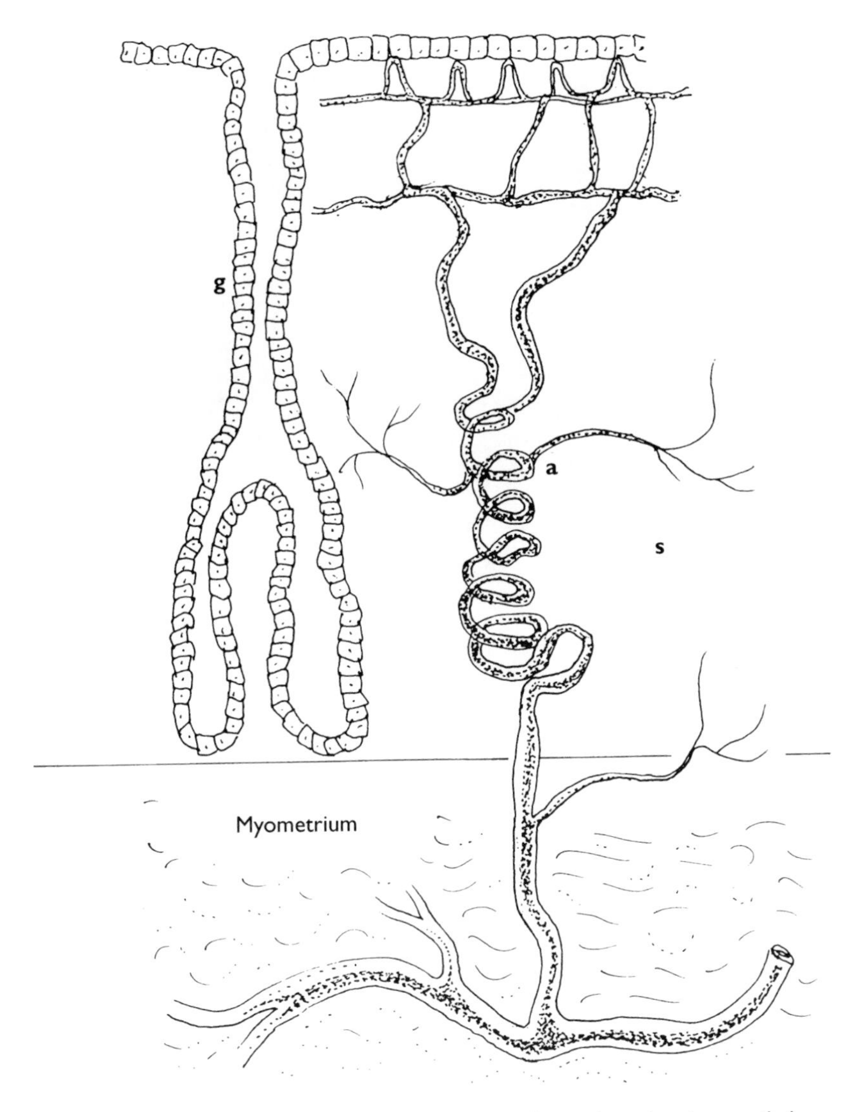

Fig. 8.2. Diagram of the vasculature of the primate endometrium showing a coiled arteriole and the superficial capillary network (after Kaiserman-Abramof & Padykula, 1989, with permission). g, gland; s, stroma; a, arteriole.

accommodate the greatly increased blood flow required by the developing fetus.

Endometrial/decidual endothelium is of use in the study of disorders of menstruation, endometriosis, endometrial neoplasia and in the development of new contraceptive agents. Failure of trophoblast migration or incomplete

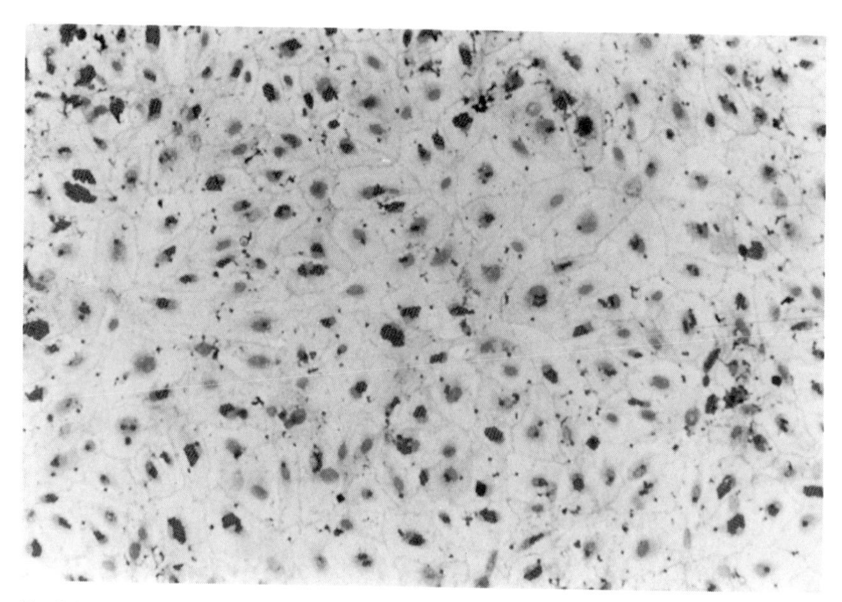

Fig. 8.3. Morphology of human decidual endothelial cells at confluence. The cells have been stained with anti-CD31. Note how the CD31 is localised to points of cell–cell contact.

transformation of maternal vessels during pregnancy are involved in the pathogenesis of pre-eclampsia, intrauterine growth retardation and still birth.

Characterisation of decidual endothelium

Human decidual endothelium displays a cobblestone morphology of flattened cells with large rounded nuclei and strong contact inhibition (Grimwood *et al.*, 1995) (Fig. 8.3). Electron microscopy reveals the presence of large striated Weibel–Palade bodies and numerous pinocytotic vesicles in the cytoplasm (Lindenbaum *et al.*, 1991).

Decidual endothelial cells express several classic endothelial markers (Table 8.1). von Willebrand factor, CD31, CD34, CD36 and lectins to which UEA-1 binds are strongly expressed on endothelial cells in tissue sections (Johannisson & Redard, 1984; Gallery *et al.*, 1991; Lindenbaum *et al.*, 1991; Grimwood *et al.*, 1995). However, CD34 and CD36 are lost when the cells are cultured *in vitro* (Grimwood *et al.*, 1995).

Table 8.1. *Reactivity of cultured decidual endothelial cells with monoclonal antibodies*

Antigen	Primary culture	First passage culture
UEA-1	+++	+++
CD31	+++	+++
von Willebrand factor	+++	+++
CD34	+	+
CD36	+	+/−
CDw90	−	−
Cytokeratin 18	−	−
Smooth muscle actin	−	−

Notes:
+, strength of immunoreactivity; −, absence of immunoreactivity.

Source of cells

The source of the cells is first trimester decidua obtained from the elective termination of healthy pregnancies. The gestational age is calculated from the first day of the last menstrual cycle and should be between 6 and 12 weeks. Decidual tissue is separated from fetal tissue by morphological examination, blood clots and mucus are aseptically removed, and the tissue then transferred to McCoy's 5A medium containing 10% FCS and antibiotics (penicillin 100 U/ml, streptomycin 100 μg/ml, gentamicin 50 μg/ml, fungizone 1 μg/ml).

Methods of isolation

Attempts at the isolation of endometrial and decidual endothelium have been undertaken since the early 1980s. Until recently, however, the various methods gave only impure cultures which were not fully characterised. In our laboratory we have undertaken extensive work on first trimester decidual endothelium which is now being extended to endometrial endothelium. Term pregnancy has been examined by Gallery *et al.* (1991), but the growth requirements of that endothelium have not been delineated.

There are six reports on the isolation and culture of decidual endothelium (Johannisson & Redard, 1984; Lindenbaum *et al.*, 1991; Drake & Loke, 1991; Gallery *et al.*, 1991; Peek *et al.*, 1994; Grimwood *et al.*, 1995). Table 8.2 lists the methods and media used by different authors. The methods fall into three groups. In the first, whole decidual explants are cultured, that is

Table 8.2. *Isolation and culture of decidual endothelial cells*

Authors	Collection medium	Isolation	Culture matrix	Culture medium	Subculture
Johannisson & Redard (1984)	199+20% FCS	No	No	199+20% FCS	No
Lindenbaum *et al.* (1991)	Milton's solution	Trypsinise	No	DMEM+10% NBCS+insulin +transferrin+sodium selenite +oestradiol+progesterone +hydrocortisone	No
Drake & Loke (1991)	199+10% FCS	QBend 40 coated beads	Gelatin	199+40% FBS+ECGS+heparin	No
Gallery *et al.* (1991)	Hank's buffer+5% FCS	UEA-1 coated beads	Gelatin	Wissler:McCoy's (1:1)+40% FCS+ECGS+heparin	No
Peek *et al.* (1994)	PBS	No	Gelatin	199+20% FCS+ECGS+heparin	No
Grimwood *et al.* (1995)	McCoy's+10% FCS	Percoll gradient and UEA-1 coated beads	Collagen type IV	McCoy's+50% HS+VEGF +magnesium+heparin +hydrocortisone+IBMX	Yes

the endothelial cells were not separated from other cells in the tissue before culture (Johannisson & Redard, 1991; Peek *et al.*, 1994). In the second, whole decidua was minced and then cultured for 10 days, following which the cultures were exposed to trypsin to remove non-endothelial cells which remained attached to the plate (Lindenbaum *et al.*, 1991). In the third, decidua was digested with collagenase and the endothelial cells separated with magnetic beads coated with the endothelial-specific monoclonal antibodies QBend 40 or UEA-1 (Drake & Loke, 1991; Gallery *et al.*, 1991). For the most part, however, these methods have proved difficult to reproduce, often giving only mixed populations of epithelial and stromal cells. Further, the endothelial cells did not proliferate in culture and could not be passaged. Our group recently developed a novel method for the routine isolation of pure endothelial cells (>95% purity) from decidua which can be passaged and maintained in culture for up to 14 days (Grimwood *et al.*, 1995). This method with the relevant chemicals (Table 8.3) will now be described in detail.

Preparation of *Ulex europaeus I* (UEA-1) coated magnetic beads

Equal volumes of UEA-1 (0.2 mg/ml in 0.5 M borate buffer, pH 9.6) and Dynabead (2×10^7/ml) are incubated overnight at 4 °C with end-over-end rotation. After coating, the beads are washed three times in phosphate-buffered saline (PBS) using a magnetic particle concentrator (MPC), resuspended at a final concentration of 2×10^6/ml in PBS/BSA, and stored at 4 °C until use.

Cell isolation and purification

Tissue is cut into 2–3 mm^2 pieces and digested with 2 mg/ml collagenase type 1A in McCoy's medium containing 10% FCS and antibiotics for 2 h with continuous stirring at 37 °C in a 5% CO_2 humidified atmosphere. The resulting cell suspension is filtered through a sterile 100 μm mesh to remove any undigested material and then washed repeatedly with PBS plus 0.1% BSA to remove collagenase. Any remaining cell clumps are dissociated by a further 5 min digestion in 0.1% trypsin containing 0.05% DNase type IV.

The endothelial cell population is initially purified using discontinuous Percoll gradient centrifugation. This was found to be essential to obtain a pure culture of endothelium from this tissue. Cells are resuspended in 20% Percoll in PBS/BSA and overlayered onto a step gradient of 40% and 60% Percoll. Centrifugation is performed at 670 *g* for 30 min. A condensed

Table 8.3. *Reagents, chemicals and equipment*

Chemicals and reagents
Chemicals, unless otherwise stated, from Sigma
Culture media and serum, unless otherwise stated, from GIBCO
Bovine serum albumin (BSA)
Collagenase type 1A
Collagen type IV (Boehringer Mannheim)
DNase type IV
Dynabeads M-450 (Dynal, UK)
Fetal calf serum (FCS)
Fucose
Fungizone
Gentamicin
Heparin
Human serum (HS, pooled male AB)
Hydrocortisone
3-isobutyl-1-methyl-xanthine (IBMX)
Magnesium sulphate
McCoy's 5A medium
Penicillin
Percoll (Pharmacia)
Phosphate-buffered saline (PBS)
Soybean trypsin inhibitor
Streptomycin
Trypsin
Ulex europeaus agglutinin-1 (UEA-1) lectin, 0.2 mg/ml in borate solution pH 9.5
Vascular endothelial growth factor (R & D Systems, UK)
Equipment
Magnetic particle concentrator (MPC; Dynal)
Mesh (nylon bolting cloth) (Lockertex, UK)

endothelial population is recovered from the 40%/60% interface. Cells are washed three times in PBS/BSA to remove Percoll by pelleting the cells at 250 *g* for 4 min and then resuspended in a small volume of McCoy's medium containing 1% FCS.

The endothelial cells are further purified using UEA-1 coated beads. Cells are mixed with the coated beads at a bead:cell ratio of 3:1 in a 0.5 ml volume and then incubated for 10 min at 4 °C with end-over-end rotation. The beads with bound cells are collected using an MPC and the bead-free supernatant with non-bound cells is removed. Attached cells are washed five times in McCoy's medium with 1% FCS. The cell–bead complex is dissociated by treatment with fucose (0.1 M) before seeding.

Establishment and maintenance of cultures

Purified endothelial cells are seeded into 35 mm^2 petri dishes, pre-coated with 5 $\mu g/cm^2$ collagen type IV. Cell proliferation is optimal when cultured on plates coated with collagen type IV or fibronectin (Grimwood *et al.*, 1995). The medium we currently use is McCoy's 5A, which has a magnesium ion concentration of 0.016 M. When the concentration of Mg^{2+} was lowered to 8 mM growth ceased.

Cells are cultured in medium supplemented with 50% human serum, VEGF (5 ng/ml), heparin (10 000 U/ml), magnesium sulphate (4 mg/ml), hydrocortisone (0.5 μg/ml), 3-isobutyl-1-methylxanthine (3.3×10^{-4}M) and antibiotics. Human serum is an absolute requirement for growth (Grimwood *et al.*, 1995). Growth increased with serum concentration to as high as 90% human serum. Such a high percentage of serum-stimulated growth was unexpected.

Medium is replaced every 2 days. The endothelial cells reach confluence after between 3 and 5 days, and may then be lifted by 5 min exposure to trypsin at 37 °C. Cells are subsequently subcultured at a split ratio of 1:2. So far, we can only passage cells for three rounds, after which the growth of the cells ceases.

Other features

The adhesion molecules ELAM, VCAM, ICAM, P-selectin (CD62) and PECAM are expressed by activated decidual endothelial cells *in vivo* (Rees *et al.*, 1993; Burrows *et al.*, 1994; Grimwood *et al.*, 1995). Expression of ELAM and VCAM is inducible by IL-1α and TNFα (Grimwood *et al.*, 1995).

Decidual endothelium, like other microvascular endothelia, has been shown to be more fastidious in the range of growth factors it responds to than large vessel endothelium. Decidual endothelium fails to proliferate in response to several known endothelial growth factors, most notably bFGF, EGF and IL-4. In addition, there was no mitogenic response of decidual endothelium using the broad spectrum endothelial cell growth supplement (ECGS), similar findings having been reported for term decidual endothelium (Gallery *et al.*, 1991). VEGF is the sole growth factor for first trimester decidual endothelial cells (Grimwood *et al.*, 1995) as yet identified. VEGF-stimulated growth was inhibited by TGFβ1 and activin.

The placenta

The human placenta is a villous haemochorial structure which is of central importance in materno-fetal transfer. It has two components: the fetal placenta (expelled at birth) and the maternal component which comprises the decidual placental bed. Fetal blood passes to the placenta via two umbilical arteries, and enters the villous capillary system where transfer of nutrients from the maternal vascular bed occurs. Blood is then returned to the fetus by the umbilical vein. Placental development plays a key role in successful pregnancy. Various angiogenic regulators, such as VEGF, FGF and PDECGF (Jackson *et al.*, 1994; Osuga *et al.*, 1995), have been implicated in placental development. There is only one report of the isolation and culture of endothelial cells from term placenta (Leach *et al.*, 1994), which follows the method described by Drake & Loke (1991) for decidual endothelial cells.

Acknowledgements

The authors are grateful to Springer-Verlag, Berlin, for permission to reproduce Fig. 8.1.

References

Bagavandoss, P. & Wilks, J.W. (1991). Isolation and characterization of microvascular endothelial cells from developing corpus luteum. *Biol. Reprod.*, **44**, 1132–9.

Burrows, T.D., King, A. & Loke, Y.W. (1994). Expression of adhesion molecules by endovascular trophoblast and decidual endothelial cells: implications for vascular invasion during implantation. *Placenta*, **15**, 21–33.

Drake, B.L. & Loke, Y.W. (1991). Isolation of endothelial cells from human first trimester decidua using immunomagnetic beads. *Hum. Reprod.* **6**, 1156–9.

Fenyves, A.M., Behrens, J. & Spanel-Borowski, K. (1993). Cultured microvascular endothelial cells (MVEC) differ in cytoskeleton, expression of cadherins and fibronectin matrix. *J. Cell Sci.*, **106**, 879–90.

Gallery, D.M., Rowe, J., Schrieber, L. & Jackson, C.J. (1991). Isolation and purification of microvascular endothelium from human decidual tissue in the late phase of pregnancy. *Am. J. Obstet. Gynecol.* **165**, 191–6.

Grimwood, J., Bicknell, R. & Rees, M.C.P. (1995). The isolation, characterisation and culture of human decidual endothelium. *Hum. Reprod.*, **10**, 101–8.

Jackson, M.R., Carney, E.W., Lye, S.J. & Knox Ritchie, J.W. (1994). Localization of two angiogenic growth factors (PDECGF and VEGF) in human placentae throughout gestation. *Placenta*, **15**, 341–53.

Johannisson, E. & Redard, M. (1984). Culture of human endothelial cells derived from capillaries of the decidual tissue. *Acta Obstet. Gynecol. Scand.*, **63**, 27–36.

Kaiserman-Abramof, I.R. & Padykula, H.A. (1989). Angiogenesis in the postovulatory primate endometrium: the coiled arteriolar system. *Anat. Rec.*, **224**, 479–89.

Leach, L., Bhasin, Y., Clark, P. & Firth, J.A. (1994). Isolation of endothelial cells from human term placental villi using immunomagnetic beads. *Placenta*, **15**, 355–64.

Lindenbaum, E.S., Langer, N. & Beach, D. (1991). Isolation and culture of human decidual capillary endothelial cells in serum-free media supplemented with human angiogenic factor. *Acta Anat.*, **140**, 273–9.

Osuga, Y., Toyoshima, H., Mitsuhashi, N. & Taketani, Y. (1995). The presence of platelet-derived endothelial cell growth factor in human endometrium and its characteristic expression during the menstrual cycle and early gestational period. *Mol. Hum. Reprod.*, **1**, 989–93.

Peek, M.J., Fraser, I.S., Johannisson, E. & Markham, R. (1994). A new reliable method of culturing and measuring human decidual capillary endothelial cells. *Exp. Toxic Pathol.*, **46**, 149–54.

Rees, M.C.P., Heryet, A.R. & Bicknell, R. (1993). Immunohistochemical properties of the endothelial cells in the human uterus during the menstrual cycle. *Hum. Reprod.*, 8, 1173–8.

Richards, R.G. & Almond, G.W. (1994). Identification and distribution of tumor necrosis factor α receptors in pig corpora lutea. *Biol. Reprod.*, **51**, 1285–92.

Spanel-Borowski, K. (1991). Diversity of ultrastructure in different phenotypes of cultured microvessel endothelial cells isolated from bovine corpus luteum. *Cell Tissue Res.*, **266**, 37–49.

Spanel-Borowski, K. & Bosch, J. van der (1990). Different phenotypes of cultured microvessel endothelial cells obtained from bovine corpus luteum. *Cell Tissue Res.*, **261**, 35–47.

Spanel-Borowski, K. & Fenyves, A. (1994). The heteromorphology of cultured microvascular endothelial cells. *Drug Res.*, **44**, 385–91.

Zhang, L., Rees, M.C.P. & Bicknell, R. (1995). The isolation and long-term culture of normal human endometrial epithelium and stroma: expression of mRNAs for angiogenic polypeptides basally and on oestrogen and progesterone challenges. *J. Cell Sci.*, **108**, 323–31.

9

Synovial microvascular endothelial cell isolation and culture

Stewart E. Abbot, Clifford R. Stevens and David R. Blake

Introduction

Contrary to popular belief the first attempts to culture endothelial cells *in vitro* pre-date considerably the work of Jaffé *et al.* (1973) or Gimbrone and others in the early 1970s. Some 50 years earlier in 1922, Warren Lewis reported the growth of endothelial cells isolated from chick liver sinusoids, and in 1925 Maximov demonstrated growth of vessels from rabbit pia. However, such cultures were essentially organ cultures which relied on high plasma concentrations to support the limited 'endothelial' like cell outgrowth from the explanted tissue. The work of Jaffé, Gimbrone and colleagues deserves mention for their pioneering work in isolating and maintaining endothelial cells in long-term culture as a single cell type, free of fibroblastic, and other non-endothelial, cell contamination.

To examine the role of particular microvascular beds in certain disease states, the long-term culture of endothelial cells derived from human microvascular beds remains a primary goal for many cell biologists. The realisation of this goal was aided immensely by the seminal articles of Judah Folkman and colleagues in the late 1970s (Folkman *et al.*, 1979) which defined the fastidious growth conditions required by certain human microvascular endothelial cells *in vitro*. Unfortunately, the laborious isolation methods employed by Folkman and Haudenschild have been less widely applicable. However, as this book highlights, there is no universal isolation and culture method which is applicable to all microvascular endothelial cell sources.

Theoretically, if a vessel can be cannulated, endothelial cells can be isolated by collagenase perfusion in a manner similar to that proposed by Jaffé for umbilical veins (Jaffe *et al.*, 1973). The application of this approach nears its limits with vessels such as coronary arteries with lumen diameters of 1–2 mm and cannot be applied to microvasculature *ex vivo*. Isolation of microvascular endothelial cells from human tissues generally requires an initial stage in which the tissue is broken down enzymatically or mechanically to yield a mixed cell population from which endothelial cells may subsequently be purified to homogeneity.

To date, two techniques have been employed to isolate microvascular endothelial cells derived from rheumatoid and osteoarthritic synovial tissue (the cells are called synovial microvascular endothelial cells; SMEC): magnetic-activated cell sorting (MACS) (Jackson *et al.*, 1990; Abbot *et al.*, 1992) and fluorescence-activated cell sorting (FACS) (Jackson *et al.*, 1989; Gerritsen *et al.*, 1993). In essence the ability of FACS to select and sort individual cells on the basis of unique cellular determinants would seem to be an ideal method by which to sort any microvascular endothelial cell type. In reality FACS-based techniques have several inherent problems, as outlined below. The following comparison is based on our experience of sorting synovial microvascular endothelial cells, and should not detract from the useful application of FACS to endothelial cells derived from other microvascular beds, detailed elsewhere in this book.

Both techniques can be used to distinguish between individual cells on the basis of cell-specific surface markers, and both techniques can, under ideal circumstances, produce exceptionally pure cultures of endothelial cells. There is, however, a high degree of cell wastage inherent in the mechanisms by which many fluorescence-activated sorters achieve high purity cell sorts. FACS methods employ a fluorescence-activated cell identification system coupled in line to an electrostatic deflection system in order to sort individual cell populations. Deflection systems cannot guarantee to sort a cell identified as 'endothelial' unless, while passing through the identification system, it is preceded and followed by other endothelial cells. If an endothelial cell is followed, or preceded, by a non-endothelial cell then all cells are rejected to ensure that the contaminating cell is not included in the sorted population. As a result of this, the use of FACS to sort primary populations of synovial cells leads to a very low yield of SMEC. FACS techniques also tend to suffer from problems associated with maintaining cell viability and sterility, but these are more easily overcome.

Despite these problems, two groups have successfully used FACS techniques, in conjunction with the ability of endothelial cells to selectively metabolise fluorochrome-labelled acetylated low density lipoprotein (diI-Ac-LDL), to isolate SMEC (Jackson *et al.*, 1989; Gerritsen *et al.*, 1993). These groups have routinely cultured mixed populations of cells, obtained from synovial tissue disaggregation, for considerable periods of time prior to endothelial cell isolation. This technique allows the propagation of sufficient numbers of endothelial cells to counterbalance the inefficiency of the sorting systems. Several factors have to be considered when maintaining endothelial cells as part of a mixed culture system. Although microvascular endothelial cell populations *in vitro* often proliferate faster when co-cultured with 'contaminating' cells, their growth rates are still considerably slower than those of contaminating fibroblasts, resulting in a lower relative proportion of endothelial cells as a percentage of the total number of cells. Additionally, it is not known to what extent the exposure of endothelial cells to protracted periods in culture may alter cell physiology from that present *in vivo*.

Modern MACS systems derive from early techniques which selectively isolated erythrocytes on the basis of their strong natural magnetic moment (conferred by their high haemoglobin content). The magnetic moment of most other cell types is too weak to permit this type of isolation; however, the magnetic moment of a cell can be artificially increased by binding magnetic particles to it. Most of the techniques which have been developed recently rely on generating an organic coating on small crystals of magnetite (Fe_3O_4) to which antibodies and other proteins, directed against specific cell surface components, may be bound (Guesdon & Avrameas, 1977). Our initial SMEC sorting protocols relied on the ability of the plant lectin *Ulex europaeus* agglutinin-1 (UEA-1) covalently coupled to small (4.5 μm diameter) magnetic particles (Dynabeads) to selectively bind to α-L-fucose residues which are present at high concentrations on endothelial cells. Certain problems were associated with this approach. Contrary to another report of a similar technique (Jackson *et al.*, 1990), we found that UEA-1 coated Dynabeads proved difficult to remove from isolated SMEC and often inhibited cell attachment to culture vessels when present at concentrations higher than 15 beads per cell. To circumvent this problem, it was necessary to estimate endothelial cell numbers accurately, prior to sorting, in order to add UEA-1 coated Dynabeads at a ratio of no more than 3 per endothelial cell. In our experience, and that of other groups (personal communication, Dynal UK), attempts to remove UEA-1 coated dynabeads from endothelial cells by addition of excess exogenous α-L-fucose were not effective.

With the development of antisera directed at constitutive endothelial cell surface antigens, a similar immunomagnetic sorting method was developed. Within synovial tissue, endothelial cells and non-proliferative macrophages are the only cell types which constitutively express the surface antigen CD31 (platelet endothelial cell adhesion molecule-1). SMEC can be isolated from mixed cultures by selective labelling with mouse anti-human CD31 antibody, prior to selection using Dynabeads coated with sheep anti-mouse immunoglobulin antibodies. The primary mixed cell populations produced from the disaggregation of rheumatoid or osteoarthritic synovial tissue specimens generally contain a few hundred thousand endothelial cells (less than 5% of the total); however, positive immunomagnetic cell selection has the ability to recover up to 90% of the cell type of interest, even when such cells exist as a small proportion of a mixed cell population (Collin-Osdoby *et al.*, 1991). Immunomagnetic cell sorting has proved to be an exceedingly efficient method with which to isolate SMEC. The technique does not adversely affect cell viability or sterility. One obvious drawback of this method is that any non-endothelial cells which are attached to positively selected SMEC will be co-sorted. Ideally, therefore, cells should be sorted from as near a single cell suspension as possible. This is achieved by careful disaggregation of the tissue and filtration of the cells prior to sorting.

Methods

The isolation method for SMEC represents the combination and development of certain features which have been reported previously for the isolation of other microvascular endothelial cells (Marks *et al.*, 1985; Dorovini-Zis *et al.*, 1991) (Fig. 9.1).

Cell source

Biopsy samples have been used as a tissue source for cultures of many human cell types; however, even if multiple biopsies are removed from the synovium the quantity of tissue is too small to yield sufficient endothelial cells to culture successfully. Fortunately, relatively large amounts of synovial tissue may be readily obtained from patients undergoing primary rheumatoid or osteoarthritic joint replacement surgery. Total hip (THR) and total knee replacements (TKR) are the most common procedures and are useful synovial tissue sources. Such tissue samples will be excised in a sterile manner and may be used immediately, or stored for limited periods in a transport medium (TM) (Hanks' Balanced Salt Solution with additions of 25 mM HEPES buffer and 3% bovine serum albumin, BSA) at 4 °C. In contrast to umbilical cords, which may be stored for considerable periods (greater than 72 h) prior to the isolation of HUVEC, the isolation of SMEC must be performed within 6–8 h post-operatively in order to maintain cell viability.

Patients undergoing TKR are subject to above-knee tourniquet prior to synovial tissue excision, and samples are therefore relatively free of blood. THR, however, are usually obtained together with substantial quantities of

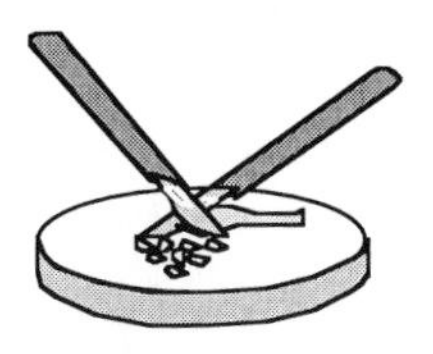

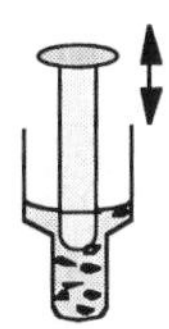

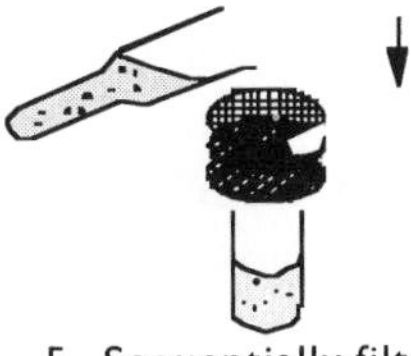

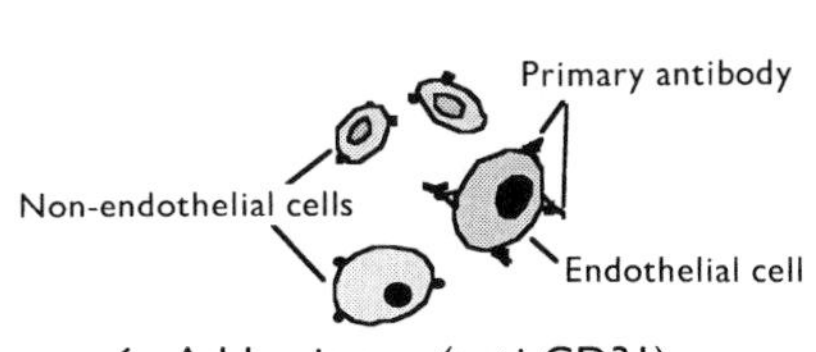

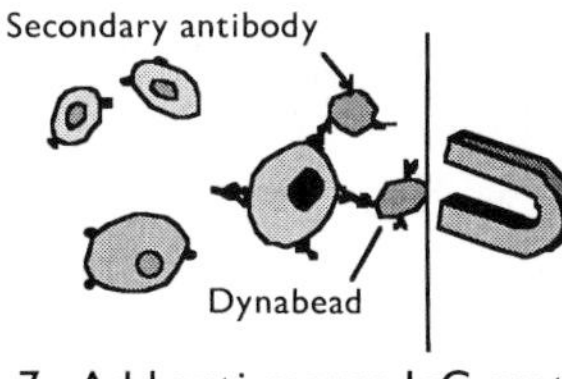

Fig. 9.1. Isolation method for synovial microvascular endothelium cells (SMEC).

blood which must be removed by washing the tissue in several changes of TM. Once washed the tissue may be inspected and the synovial membrane dissected clear of subsynovial tissue. A dissecting microscope aids this and subsequent steps.

Synovial membrane hyperplasia is a common feature of rheumatoid and occasionally osteoarthritic synovitis, with the hyperplastic tissue appearing as highly vascularised, villus-like protrusions from the synovial membrane. This tissue is an ideal source of cells for three reasons. First, hyperplastic villi are readily removed from the sub-synovium by immersing the tissue in TM and allowing the villi to float upwards, before cutting through the base of the villi using springbow forceps; secondly, villous tissue contains relatively few large vessels; and, lastly, the microvasculature is in a process of active angiogenesis, which may facilitate the initial propagation of the cells *in vitro*. Once dissected the synovial tissue may be further disaggregated by enzymatic methods.

It may be argued that SMEC isolated from hyperplastic villi represent a subpopulation of cells which may be atypical of other microvasculature present within synovial membrane tissue. In our experience SMEC display a number of characteristics which differ in comparison with endothelial cells isolated from umbilical veins (Abbot *et al.*, 1992) or the microvasculature of human subcutaneous adipose tissue (unpublished results); however, little variation is observed between endothelial cells isolated from different regions of rheumatoid synovium.

For experimental purposes it is highly desirable to obtain non-diseased 'normal' synovial tissue as an appropriate control; however, this poses many problems. Normal tissue may occasionally be obtained from limb amputation following traumatic injury, but such cases are few and unscheduled. Normal tissue may also be obtained from distal joints in limbs undergoing amputation for malignancy, but such cases are also rare and usually only conducted at specialist centres. As a final confounding problem, while the hyperplastic intimal synovial layer of an inflamed joint may be hundreds of cells in thickness, normal synovium is usually only 60–100 μm thick, making dissection difficult and yielding somewhat fewer endothelial cells.

Tissue disaggregation

With the exception of certain explant culturing techniques, most tissues must be disaggregated prior to placing cells into culture. Methods of tissue disaggregation for animal cell culture vary; however, many rely on a mixture of mechanical and enzymatic processes. Mechanical chopping of the tissue is initially performed in order to increase the efficiency of subsequent, diffusion-limited, enzymatic processing.

In the case of inflamed rheumatoid synovial tissue, synovial membrane tissue is repeatedly chopped with crossed scalpel blades to yield tissue fragments of approximately 3 mm^3. Alternatively, as mentioned earlier, the vascularised villus-like protrusions (1–6 mm in length) may be trimmed from the tissue. It is at this stage that the microvascular nature of the cells to be isolated is confirmed. Using a dissecting microscope, tissue pieces containing vessels with a lumen diameter greater than approximately 150 μm are discarded. This procedure is less time consuming when villous material is used as it generally contains relatively few vessels greater than 150 μm in diameter. The remaining tissue is then washed, vortex mixed and the tissue pieces allowed to sediment under unit gravity for 3 min before being removed for further enzymatic breakdown. This process facilitates the removal of cells damaged during the mincing processes.

Enzymatic breakdown of the tissue is directed at the dominant components within the basement membrane material of the tissue, namely type IV collagen and hyaluronic acid. The enzymatic treatment has to be sufficient to reduce the tissue fragments to as near to a single cell suspension as possible without unduly compromising the viability or plating efficiency of the endothelial cells once isolated. This process is carried out for 60–90 min at 37 °C using a mixture of 0.2% type IV collagenase (CLS4, Worthington) and 0.05% bovine testes hyaluronidase in TM. Worthington collagenase is used as it has very low tryptic activity which would otherwise decrease cell viability during prolonged incubation.

Following incubation, the tissue pieces are gently homogenised (10–15 slow strokes) using a loose-fitting siliconised* glass/PTFE homogeniser. The resulting cell suspension is vortex-mixed before being passed through a steel 250 μm pore filter (to remove any large tissue pieces) followed by 110 and 55 μm nylon filters (to remove cell clumps). While in reality an absolute single cell suspension may be difficult to achieve, the non-specific interaction of magnetic beads with aggregated cell clumps is one of the major factors in limiting the purity of the final cell population (Fig. 9.2).

Immunomagnetic sorting

To help disaggregate small cell aggregates which may escape filtration, the final filtrate may be repeat pipetted through a siliconised flame-polished glass pipette. At this stage, total cell count should be performed, endothelial cell numbers being calculated by assuming 3–5% of the total. If problems are encountered with cell viability, it is a good idea to check total cell viability by trypan blue exclusion at this, and subsequent stages in the isolation.

All the subsequent steps should be conducted on ice at 4 °C, initially to prevent antigen 'capping' and subsequently to prevent phagocytosis of the Dynabeads by non-endothelial cells.

* To reduce cell to substrate adhesion it is preferable, where practical, to siliconise all equipment used during the isolation procedure which is not maintained at 4 °C. Siliconisation is simply performed by immersing or filling equipment with a siliconisation fluid (Sigmacote, Sigma) for 5 min before removing excess fluid and air drying. Siliconised equipment should be rinsed in double distilled water immediately prior to use.

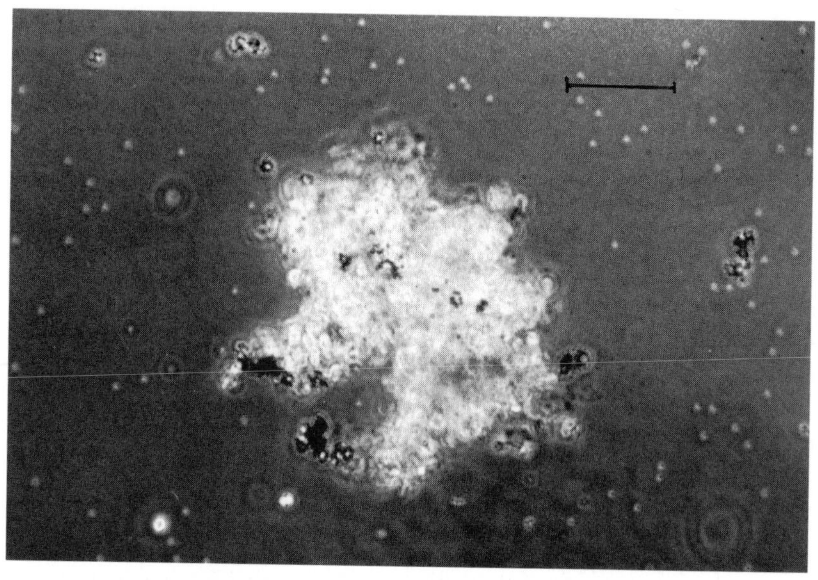

Fig. 9.2. A cell aggregate which has escaped filtration; Dynabeads are either associated with endothelial cells within the aggregate or non-specifically trapped between the cells. Non-endothelial cells isolated in this manner will tend to overgrow endothelial cultures within a few weeks. Scale bar represents 50 μm.

The cells are sedimented by centrifugation (400 *g*/5 min) before being resuspended in 500–1000 μl of PBS/1% BSA with the addition of 1 μl of anti-CD31 antibody per 10^6 endothelial cells (1 mg/ml mouse IgG_1; R&D Systems, Oxford) for 30 min on ice. Following incubation, the unbound antibody is removed by washing the cells three times in PBS/1% BSA. The antibody-tagged endothelial cells are isolated by the addition of sheep anti-mouse IgG_1 antibody-coated paramagnetic Dynabeads (Dynal, Sweden). We have found that the isolation procedure is effective when beads are added at a ratio of 3:1 (beads:endothelial cells) for 20 min in 500 μl of PBS/1% BSA at 4 °C. Endothelial cells bound to Dynabeads are isolated by the application of a permanent magnetic field* to the side of the isolation vessel while non-bound cells are removed by aspiration (in the first instance, this aspirate is retained for further rounds of endothelial cell isolation). The magnetic field is removed and the endothelial cells resuspended in 1–2 ml of PBS/1% BSA

* Dynal recommend the use of a Magnetic Particle Concentrator to isolate cell–bead complexes; however, any strong permanent magnetic field which can be placed in close proximity to the isolating vessel should be equally effective. Initially, if the cell density is high, endothelial cells may take up to 3 min to migrate towards the magnet; however, during subsequent washing steps cells will be retained by the magnetic field in less than 30 s.

before re-applying the field. The aspirate is removed and discarded and this process repeated three times to ensure the removal of all non-endothelial cells. Anti-CD31 tagged endothelial cells exhibit various affinities towards the Dynabeads, with some cells being isolated with only one attached bead while others may have 5–10. In many cases addition of more Dynabeads and repetition of the procedure will yield further endothelial cells which failed to adhere to the beads during the first round of isolation. Re-addition of Dynabeads to the initial aspirate is conducted assuming an endothelial cell proportion of 1% of the total.

If a high level (greater than 20%) of fibroblastic cell contamination is observed within primary cultures, endothelial cells can be re-sorted by removing them by scraping and repeating the above procedure using a suitable estimate of endothelial cell number. If re-sorting is to be effective it must be conducted within 7 days of establishing the culture, as isolated SMEC gradually lose constitutive CD31 expression within 7–14 days in culture.

Isolated cells may be cultured directly or, in cases when bead:cell ratios of greater than 10:1 are observed, beads may be removed using the Dynal product Detach-a-bead. Detach-a-bead is a product developed specifically to facilitate the removal of $CD4^+/CD8^+$ T lymphocytes from Dynabeads following positive selection; however, it has proved of some use in dissociation of beads from endothelial cells isolated by the above protocol.

The isolated endothelial cells are centrifuged (400 *g*/5 min) before resuspension in 100 μl of PBS/1% BSA containing 10 μl of Detach-a-bead per 10^7 cells. This solution is vigorously vortex-mixed for 10 s before incubation at room temperature for 30 min with occasional mixing by inversion. After treatment, endothelial cells and unattached beads are retained in the magnetic field while excess Detach-a-bead is removed by aspiration. Detach-a-bead treatment seldom removes all the bound beads but should consistently remove sufficient beads to preserve the cell plating efficiencies, which are adversely affected by bead:cell ratios of greater than approximately 15:1. Detach-a-bead appears to act by inhibiting the reformation of antibody bonds mechanically broken by vortexing.

As a caveat to the use of Detach-a-bead, many endothelial cells fail to attach to the culture vessels even after the removal of a large proportion of their attached Dynabeads, probably due to the loss of viability induced by the vigorous nature of the vortexing required to break the bonds formed between cells and beads.

Cells which are cultured directly usually shed the majority of attached beads after 5 days in culture. Isolated SMEC are sensitive to seeding density and will tend to grow slowly unless seeded at a density in excess of 1000 cells/cm^2.

Cell isolation protocol

1 Obtain synovial membrane tissue from tissue excised from hip or knee joints during rheumatoid-arthritic joint replacement surgery.
2 Place the excised tissue immediately in transport medium (Hank's Balanced Salt Solution with additions of 25 mM HEPES buffer and 3% bovine serum albumin (BSA)), and store at 4 °C until processed.
3 Wash the tissue in several changes of transport medium to remove excess blood from the sample prior to dissecting the synovial membrane free of subsynovium and capsular material.
4 Mince the remaining tissue into approximately 3 mm^3 fragments, using crossed scalpels.
5 With the aid of a dissecting microscope, determine the microvascular nature of the cells to be isolated by discarding tissue fragments containing vessels with lumen diameters greater than approximately 150 μm.
6 Wash the remaining tissue in TM and vortex mix to help dissociate cells which have been damaged during the tissue mincing. Following washing, the tissue fragments can be collected by sedimentation under unit gravity for 3 min.
7 Incubate the tissue fragments in 0.2% type IV collagenase (CLS4, Worthington)/0.05% bovine testes hyaluronidase in TM for 60–90 min at 37 °C.
8 Gently homogenise the tissue (10–15 slow strokes) using a loose-fitting siliconised glass/PTFE homogeniser before vortex mixing the resulting cell suspension.
9 Sequentially filter the cell suspension through a steel 250 μm pore filter (to remove any large tissue pieces) followed by 110 and 55 μm nylon filters (to remove aggregated cells). Any remaining cell clumps may be dissociated by a further 15 min incubation in fresh 0.2% type IV collagenase before being repeat pipetted through a flame polished, siliconised glass pasteur pipette.
10 Count the cells using a haemocytometer. Assume the endothelial cell numbers are 3–5% of the total.
11 Resuspend the cells in 500 μl of PBS/1% BSA (or heat-inactivated FCS) prior to the addition of 1 μl of mouse IgG_1 anti-human CD31 (1 mg/ml) per 1×10^6 endothelial cells.
12 Incubate the antibody/cell suspension at 4 °C for 30 min, with occasional mixing by inversion. Pellet the cells by centrifugation (400 *g*, 5

min), then aspirate the unbound antibody. Wash the cells in 10 ml of PBS/1% BSA and repeat the centrifugation/aspiration.

13 Add anti-mouse IgG_1 coated paramagnetic Dynabeads (3:1 ratio of beads to endothelial cells) in a minimal volume (~500 μl PBS/1% BSA) for 20 min at 4 °C.

14 Retain endothelial cells bound to Dynabeads in a magnetic field while non-bound cells are removed by aspiration.

15 Retain the aspirate and repeat steps 11–14 assuming an endothelial cell proportion of 1% of the total.

16 If the bead to cell ratio exceeds 15:1, centrifuge (400 *g*, 5 min) the isolated cells and resuspend in 100 μl of PBS/1% BSA containing 10 μl Detach-a-bead per 1×10^7 cells. Vigorously vortex mix this solution for 10 s before incubating at room temperature for 30 min with occasional mixing by inversion.

17 Pool isolated endothelial cells, count and seed onto fibronectin (5 $\mu g/cm^2$) coated culture dishes at a density of ~2000 $cell/cm^2$ in medium 199 with additions of 15% v/v heat inactivated fetal calf serum (FCS), 15% v/v heat inactivated human serum, 50 μg/ml endothelial cell growth supplement (ECGS), 20 ng/ml epidermal growth factor (EGF), 30 IU/ml heparin, 50 IU/ml penicillin G and 50 μg/ml streptomycin sulphate.

18 When confluent, subculture cells with a split ratio of no greater than 3:1.

Culture conditions

In 1981 Bruce Zetter made the prediction that 'it would be very surprising if, in another two years, anyone were still culturing these (microvascular) cells in high serum or tumour conditioned medium'. Some 15 years later we have made some progress away from the use of tumour conditioned medium; however, the development of a universal microvascular endothelial cell culture medium is still eagerly awaited.

Isolated SMEC can be grown in culture vessels coated with either gelatin (1% w/v bovine skin gelatin (225 bloom) in PBS, overnight at 4 °C) or fibronectin (5 $\mu g/cm^2$ in double distilled water for 45 min prior to air drying vessel); however, primary cultures tend to grow better on fibronectin coated plastic. Cells may be subcultured subsequently onto gelatin coated vessels without noticeable deterioration in growth rates. Surprisingly, culture vessels coated with laminin or type IV collagen are particularly poor at sustaining cell growth. Isolated SMEC are routinely cultured in medium 199 with additions as detailed in point 17 above. Certain cultures may occasionally exhibit poor initial growth rates, often such cases respond favourably to increasing the concentration of human serum to 30% v/v until the cells are subcultured for the first time. SMEC are routinely sub-cultured using a split ratio of 3:1. The use of animal and human sera to enhance cell growth rates *in vitro* is subject to problems with regard to batch variation between different preparations. We have found that certain batches of human serum are particularly poor at supporting SMEC growth. While there is unfortunately no substitute to a trial and error approach in testing different sera batches, once a suitable batch of sera has been identified, many manufactures will reserve quantities for your continued use.

Low levels of fibroblastic contamination (up to 3%) may be removed by manual 'weeding' with the aid of a sterile, flame-drawn, beaded haematocrit tube and phase contrast microscope. Alternatively, limited trypsinisation of cell populations at subculture (transfering the cells when approximately 80% of the cells have detached) tends selectively to remove endothelial cells while the relatively trypsin-insensitive fibroblastic cell contamination remain attached.

We have maintained isolated SMEC in culture for up to nine subcultures, and while similar cell morphology is observed throughout, subtle changes appear after four or five subcultures. As SMEC are subcultured they tend to adopt a slightly larger spread area with the cytoskeleton becoming more apparent when viewed using phase optics. However, with the exception of the occasional migrating cell, cells always adopt strict contact-inhibited monolayers at confluence. Additionally, as SMEC are subcultured the growth requirements of the cells become somewhat less stringent, possibly reflecting the expansion of a sub-population of endothelial cells with less demanding growth requirements.

Cryopreservation

No special problems are encountered with the cryopreservation of SMEC, except to mention that cells are not particularly hardy and are best frozen from subconfluent cultures using a good controlled rate freezing vessel (−1 °C/min). Cultures are best re-established using high initial seeding densities (greater than 2000 cells/cm^2). SMEC may be stored frozen in liquid nitrogen for periods of up 1 year in medium 199 with additions of 10% heat inactivated FCS and 10% dimethylsulphoxide (1×10^6 cells/ml). Dimethylsulphoxide must be removed rapidly when vials are thawed.

Cell characterisation

Together with their ability specifically to isolate endothelial cells from umbilical vein Jaffé and colleagues advanced endothelial cell biology by the introduction of methods to assess the lineage of putative endothelial cells in culture. Their identification of endothelial cells was based on several criteria: a distinctive 'cobblestone' morphology, presence of Weibel–Palade bodies, ABH antigens and the presence of an anti-haemophilic factor or von Willebrand factor (vWF). Although cobblestone morphology and immunohistochemical localisation of vWF are still widely used in the characterisation of large vessel endothelial cells, the demonstration that these criteria are not unique to endothelial cells (Chung-Welch *et al.*, 1989) has necessitated the adoption of additional characterisation methods for isolated microvascular endothelial cells.

With the development of numerous anti-sera and the identification of plant lectins which selectively bind to endothelial cell determinants, the list of possible characterisation procedures is extensive. Here the list is limited to those markers which positively identify SMEC and which are easily applied on a routine basis. The characterisation procedures include morphology, immunohistochemical localisation of vWF, selective metabolism of diI-Ac-LDL together with substrate-induced differentiation.

Large vessel endothelial cells are routinely isolated as sheets of cells containing greater than a hundred cells. In culture these cells rapidly adopt a typical cobblestone morphology. In contrast, however, many microvascular-derived endothelial cells, including SMEC, are placed in primary culture either singly or as small groups of two to ten cells. On suitable substrates these single cells spread and adopt a round 'fried egg' morphology, and form cobblestone-type patterns only on contact with other endothelial cells (Fig. 9.3). Occasionally, isolated SMEC display an elongated morphology intermediate between the 'fried egg' pattern and the bipolar morphology typical of fibroblasts in culture; however, as stated earlier, this morphology reverts to a cobblestone-type pattern as the culture approaches confluence. It has been reported that microvascular endothelial cells derived from bovine corpus luteum display a range of morphologies *in vitro* (Spanel-Borowski & van der Bosch, 1990) such a range of morphologies is not observed in SMEC cultures.

Procedures for the immunohistochemical localisation of vWF are routine and will not be dealt with in depth here, except to mention that cells to be stained are usually grown to confluence on acetone-resistant Thermanox or fibronectin coated (5 $\mu g/cm^2$) glass coverslips fixed by immersion in 1:1 cold acetone:methanol for 10 min. This fixation method facilitates the intracellular localisation of vWF (Fig. 9.4).

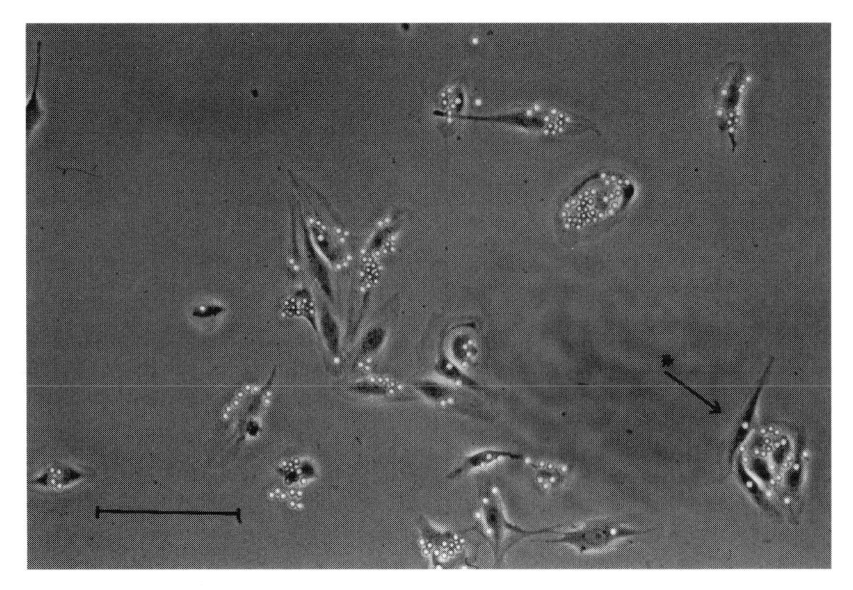

Fig. 9.3. Freshly isolated SMEC; cells tend to form small groups of 2–10 cells (asterisked arrow); beads are still present on these cells but will gradually detach as cells are maintained in culture. Scale bar represents 50 μm.

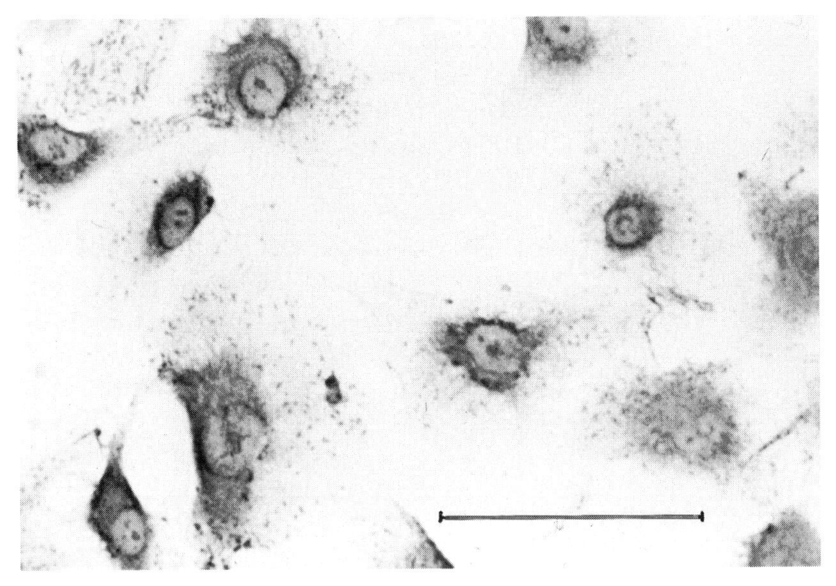

Fig. 9.4. Immunohistochemical localisation of von Willebrand factor on passage 3 SMEC; note the positive discrete cytoplasmic granular staining pattern. Scale bar represents 50 μm.

A somewhat more specific marker of endothelial and non-proliferative macrophage cells is their ability to metabolise fluorescently labelled acetylated low density lipoprotein via a 'scavenger' LDL receptor (Voyta *et al.*, 1984). Cells cultured in the presence of 1,1′dioctadecyl-3-3,3′,3′-tetramethylindocarbocyanine perchlorate (diI)-acetylated low density lipoprotein (diI-Ac-LDL) (10 μg/ml; Collaborative Research) for 4 h display a fluorescent granular pattern when viewed via 514 nm excitation and 550 nm emission filters. This technique has the advantage that it can be performed on live cultures, using a suitable inverted fluorescence microscope permitting cell subculture.

Currently, the single most discriminative marker of a putative endothelial cell lineage is their ability to differentiate on suitable extracelluar matrix substitutes. McGuire & Orkin (1987) observed that Matrigel, a solubilised basement membrane preparation extracted from the Engelbreth–Holm–Swarm mouse sarcoma, induces the rapid differentiation of monolayer cultures of endothelial cells to form 'vessel-like' tubes *in vitro*. Endothelial cells are passaged onto a thin film of Matrigel* (5000 cell/cm^2) in normal culture medium before being incubated for 4–8 h. During this time endothelial cells differentiate to form phase bright vessels (Fig. 9.5).

The isolation procedure itself provides a further confirmation of the endothelial nature of isolated cells. While CD31 expression may be observed on other cell types it is not expressed by the most common culture contaminant: fibroblasts. However, as stated earlier, constitutive SMEC CD31 expression slowly disappears over a period of 7–14 days in culture. Further, less routine SMEC characterisation methods include electron microscopic localisation of Weibel–Palade bodies, immunohistochemical localisation of angiotensin converting enzyme, cytokine-induced E-selectin expression and recognition of surface α-L-fucose by the plant lectin UEA-1.

Experimental use and further work

The synovial microvascular endothelial cells which we have isolated and cultured, using the methods detailed above, have been used extensively to investigate the expression of cellular adhesion molecules and processes regulating leucocyte margination in rheumatoid synovitis. It has become apparent that, with regard to certain aspects of adhesion molecule expression and leucocyte adhesion, isolated SMEC represent an endothelial cell population distinct from human endothelial cells isolated from umbilical vein or subcutaneous

* Matrigel must be stored and handled at less than 22 °C as above this temperature it spontaneously forms a gel.

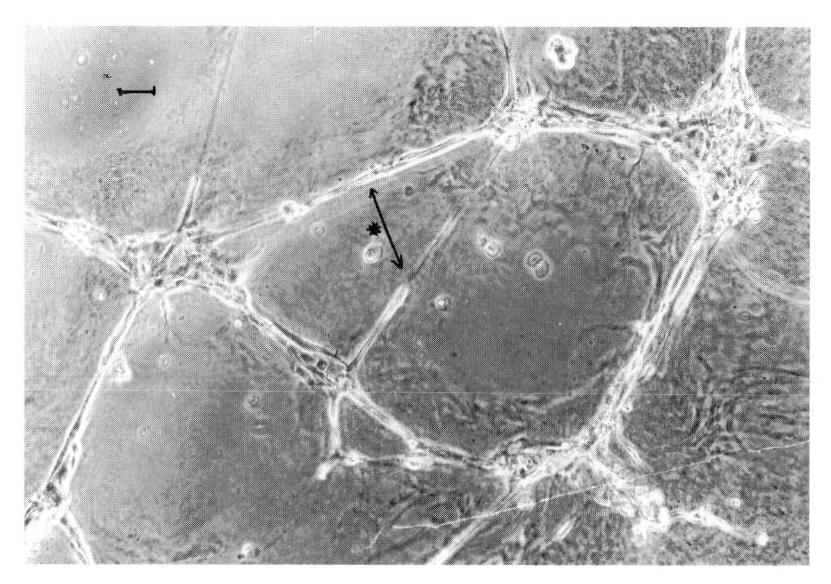

Fig. 9.5. *In vitro* differentiation of passage 3 SMEC seeded onto Matrigel coated culture vessel for 6 h; note the presence of phase bright 'vessels' running between groups of confluent cells (asterisked arrow). Scale bar represents 50 μm.

adipose tissue. While SMEC are exquisitely sensitive to the induction of E-selectin by the pro-inflammatory cytokines interleukin-1 or tumour necrosis factor, these cytokines largely fail to induce vascular cell adhesion molecule-1 expression in these cells. The expression of E-selectin correlates with the *in vitro* adhesion of CD45RO+ T lymphocytes and neutrophils to SMEC when assessed by dynamic adhesion assay techniques, and may help explain patterns of leucocyte margination observed *in vivo*.

We have successfully applied the immunomagnetic sorting methods detailed above to the isolation of microvascular endothelial cells derived from human subcutaneous adipose tissue and decidual tissue without problem. However, problems have been encountered in developing media capable of sustaining decidual microvascular endothelial cell growth *in vitro* (see Chapter 8). Future developments of the techniques will centre around the need to further define which elements of fetal calf and human sera are necessary for the propagation of human microvascular endothelial cells *in vitro*.

Acknowledgements

We would like to acknowledge the help of the following: Mrs Margaret West and Shirley Crosbie for their help in the collection of tissue samples which were kindly provided by Dr Gareth Scott and Mr Michael Freeman; Roy Edwards (Product Manager, Dynal UK) for technical advice; and the Arthritis and Rheumatism Council for helping to fund the development of the methods.

References

Abbot, S.E., Kaul, A., Stevens, C.R. & Blake, D.R. (1992). Isolation and culture of synovial microvascular endothelial cells: characterisation and assessment of adhesion molecule expression. *Arthritis Rheum.*, **35**, 401–6.

Chung-Welch, N., Patton, W.F., Yen-Patton, G.P.A., Hechtman, H. & Shepro, D. (1989). Phenotypic comparison between mesothelial and microvascular cell lineages using conventional endothelial cell markers, cytoskeletal protein markers and *in vitro* assays of angiogenic potential. *Differentiation*, **42**, 44–52.

Collin-Osdoby, P., Gursler, M.J., Webber, D. & Osdoby, P. (1991). Osteoclast-specific monoclonal antibodies coupled to magnetic beads provide a rapid and efficient method of purifying avian osteoclasts. *J. Bone. Miner. Res.*, **6**, 1353–65.

Dorovini-Zis, K., Prameya, R. & Bowman, P.D. (1991). Culture and characterization of microvascular endothelial cells derived from human brain. *Lab. Invest.*, **64**, 425–36.

Folkman, J., Haudenschild, C.C. & Zetter, B.R. (1979). Long term culture of capillary endothelial cells. *Proc. Natl. Acad. Sci. USA*, **76**, 5217–21.

Gerritsen, M.E., Kelly, K.A., Ligon, G., Perry, C.A., Chien-Ping, S., Szczepanski, A. & Carley, W.W. (1993). Regulation of the expression of intercellular adhesion molecule-1 in cultured human endothelial cells derived from rheumatoid synovium. *Arthritis Rheum.*, **36**, 593–602.

Guesdon, J.L. & Avrameas, S. (1977). Magnetic solid phase enzyme-immunoassay. *Immuno-chemistry*, **14**, 443–7.

Jackson, C.J., Garbett, P.K. & Marks, R.M. (1989). Isolation and propagation of endothelial cells derived from rheumatoid synovial microvasculature. *Ann. Rheum. Dis.*, **48**, 733–6.

Jackson, C.J., Garbett, B., Nissen, B. & Schrieber, L. (1990). Binding of human endothelium to *Ulex europaeus*-1 coated Dynabeads: application to the isolation of microvascular endothelium. *J. Cell Sci.*, **96**, 257–62.

Jaffé, E.A., Nachman, R.L., Becker, C.G. & Minick, C.R. (1973). Culture of human endothelial cells derived from umbilical veins. *J. Clin. Invest.*, **52**, 2745–51.

Marks, R.M., Czerniecki, M. & Penny, R. (1985). Human microvascular endothelial cells: an improved method for tissue culture and a description of some singular properties in culture. *In Vitro Cell Dev. Biol.*, **21**, 627–35.

McGuire, P.G. & Orkin, R.W. (1987). Isolation of rat aortic endothelial cells by primary explant techniques and their phenotypic modulation by defined substrata. *Lab. Invest.*, **57**, 94–105.

Spanel-Borowski, K. & Bosch, J. van der (1990). Different phenotypes of cultured microvessel endothelial cells obtained from bovine corpus luteum. *Cell Tissue Res.*, **261**, 35–47.

Voyta, J.C., Via, D.P., Butterfield, C.E. & Zetter, B.R. (1984). Identification and isolation of endothelial cells based on their increased uptake of acetylated low density lipoprotein. *J. Cell Biol.*, **99**, 2034–40.

Zetter, B.R. (1981). The endothelial cells of large and small blood vessels. *Diabetes*, **30**, 24–8.

Index

Page numbers in *italic* type refer to illustrations.